Mapping the
Human Genome

Mapping the Human Genome

Reality, Morality, and Deity

Theodore C. Kent

UNISERSITY
PRESS OF
AMERICA

Lanham • New York • London

Copyright © 1995 by
University Press of America® Inc.
4720 Boston Way
Lanham, Maryland 20706

3 Henrietta Street
London WC2E 8LU England

Library of Congress Cataloging-in-Publication Data

Kent, Theodore C.
Mapping the human genome : reality, morality, and Deity /
Theodore C. Kent.
p.   cm.
Includes bibliographical references.
1. Gene mapping—Moral and ethical aspects.   2. Gene mapping—
Psychological aspects.   I. Title.
QH445.2.K46   1995   174'.25—dc20   94–35665 CIP

ISBN 0–8191–9784–X (cloth : alk. paper)
ISBN 0–8191–9779–3 (pbk. : alk. paper)

The paper used in this publication meets the minimum requirements of
American National Standard for Information Sciences—Permanence
of Paper for Printed Library Materials, ANSI Z39.48–1984.

to my wife, Shirley

and my daughter, Susan

TABLE OF CONTENTS

FOREWORD

Certain dates in history, such as 1492, stand out as reminders of the dawn of an era. Future generations may look on the year 1953 as a date to remember, since in that year, two men accomplished the first successful climb of Mount Everest and another two men reported the molecular structure of DNA. The first of these accomplishments had an immediate impact as a great example of human initiative and endurance. On the other hand, few people outside the biomedical research community attached the same significance to the scientific advance that launched a new scientific discipline, molecular biology. It might be said that 1953 was the year in which began the age of the genome. It was a time in which power of modern science created great fears of global destruction, compelling humanists as well scientists to face the moral consequences of their discoveries. In fear of Armageddon, scientists, politicians, scholars, and concerned human beings in nations around the world struggled to prevent the kind of conflict that would end this era. In doing so they have initiated discussion of the moral consequences of scientific discovery. Any such discussion requires an effort to master the basics of scientific colloquy and deep immersion in it's literature. Few are up to the task. Dr. Theodore Kent has devoted a long and unceasingly productive life to the study of human psychology both as a scientist and as a man vitally and passionately concerned with the broad moral issues of our time. More than five books and scores of articles represent his contribution to human discourse and his new work "Mapping the Human Genome" speaks to our concerns about the unmapped powers of genetic engineering while remembering the lesson of Job; virtue must be pursued for it's own sake.

Henry C. Powell, M.D., M.R.C.Path.
Professor of Pathology (Neuropathology)
School of Medicine,
University of California, San Diego

PREFACE

Mapping the human genome entails locating each gene on a molecule and matching it with its specific functions. Theoretically scientists using a map that locates the position of every gene in the human genome would have the ability to redesign human beings by replacing specific genes with other genes. Although genetic engineering remains a vastly complicated procedure, in principle it would not be altogether different from how my young grandson builds and rebuilds towers when he plays with his blocks.

The busy gene detectives in their crowded laboratories have already located genes that cause some of our most serious diseases; such as Huntington's disease, amyotrophic lateral sclerosis (better known as Lou Gehrig's disease), Parkinson's disease, and cystic fibrosis. The list seems to grow longer every few months. Recently genes believed to play a role in some personality disorders have been identified—an idea that makes some of us uncomfortable because we have become accustomed to considering the environment as the major influence on personality. This book examines the psychological and ethical problems that genetic engineering may precipitate. It explores human behavior in the light of factors that have shaped its development in the modern world including our perceptions of morality, religious values, and historical developments.

Opinions for and against the use of genetic methods to change aspects of human life will be shaped by peoples' perceptions of reality, morality, and deity. These are the concerns of religion, philosophy and, to a growing extent, of science, especially the behavioral sciences. Traditional social and religious beliefs incorporate the fundamentals of various faiths. Religious and other institutions serve the human need for stability and give answers to questions that science cannot provide. Their impact on the future application of gene technology will be discussed with this in mind.

In the pages that follow, we explore the consequences of genetic engineering from the point of view of a behavioral scientist. We shall discuss the roles that values, concepts of reality, and our relationship to the universe play in the Age of the Genome. The journey undertaken by genetic engineering that began in this century is hazardous with unpredictable results that may have potentially damaging, irreversible consequences.

As we approach a sharp turn in the history of our civilization, it is possible that situational and utilitarian values will emerge to replace those that have sustained humankind through the millennia. The goal of the book is to help us prepare ourselves psychologically and emotionally to cope with the biotechnical innovations genetic engineering may inject into our lives. Throughout, we shall view gene technology from its broadest dimensions; for it will effect not only ourselves and our globe, but also, the very nature of the universe itself. Therefore, the dynamics of the universe, the evolutionary pressures on life, and genetic engineering are examined in the light of their interdependence.

We shall face unprecedented challenges. In the light of these challenges we will find it worthwhile to reexamine the present ideas of human nature and explore some new ones. I ask the reader to maintain an open mind. Behavioral science has usually excluded abstract ideas such as reality, morality, and deity from its agenda. This has curtailed its capacity to provide insights into the human aspects of life and existence. However, in considering the impact of gene technology on human life, only the dimensions of the Big Picture, the human-cosmos relationship, can reveal the full meaning of its impact on the future of our species. The small picture view confines itself to our species and even more minutely, to the survival of our gene pool as if this were the only legitimate concern. It sees our spiritual nature as derived from our neuroanatomy. This book attempts to convince the reader that the narrow approach reverses the logical order of things and ends in distortion. Here we look beyond this narrow preoccupation and pay heed to our relationship with nature as a whole.

The dynamics of the universe flowing through us could change the way we live and think about the world and our place in it. Both the philosopher Nietzsche and spiritual teachers like historical Jesus envisioned a future of changed human values. As one can imagine, their perceptions differed widely. In the chapter before the final one, the reader will be given the opportunity to consider which changes are appropriate for the Age of the Genome. The different points of view will be presented in the final chapter leading to my conclusions regarding the changes that will become necessary if the mapping of the human genome will ultimately benefit and not destroy humankind.

Acknowledgments

I thank my wife, Shirley, and my daughter, Susan, for their support and encouragement. My wife edited the book and made valuable suggestions throughout. I am grateful to Twilla McGlone who came to my rescue by correcting errors on the computer and also gave advice on style. Dorethea Mallow, a writer herself, did much of the proofreading and made suggestions. Some friends and colleagues offered helpful comments on several of chapters. I thank Darryl and Larry DiRuscio for their assistance in the manuscript's preparation.

INTRODUCTION

Ahoy, you who set sail in fragile barque!
In eagerness to hear my message,
You follow in wake of my sail,
Humming as it plows the sea.
Look back and reflect upon
Your own familiar shores.
Do not commit yourselves
To sail the seas I sail
Lest, loosing sight of me, you founder.
The uncharted waters I traverse,
Have never been crossed before.[1]

Thus does Dante warn us to think twice before we enter
"Paradise." There is a new paradise being offered us in the
form of a panacea designed to remedy all our ills and even,
eventually, to transcend our limitations. It offers us the hope
and promise that we will, in the future, be able to create the
"new human being" and the "new world" in the image of our
choice. Indeed, the stunning successes of modern biological
sciences in medicine, industry, agriculture, and especially

---

[1]"Il Paradiso" (II-6), *L'Opere di Dante: Testo Critico*. (1960) Florence: Societa
Dantesca Italiana. (My translation.)

genetics, has made us very optimistic about the future possibilities of realizing this dream. *But at what cost?* What are the implications of pursuing this dream? And if we attain it, what is it about ourselves as human beings that we risk loosing?

The chapters that follow are, among other things, an attempt to get us to think now about such important issues. Mankind has always had utopian aspirations. We are driven by the desire to solve, once-and-for-all, the serious problems in life and to create our own Eden. And our latest version of the "original temptation" comes in this form: if only we can let the genome out of the bottle, we will indeed "be like gods" (Genesis: 3:5).

But even if this were possible, would it be desirable? What would it mean in regards to the richness of human diversity, with our being creatively in the world, and, in turn, being "created" by the world we are in? The Spanish philosopher, José Ortega y Gasset, reminds us that we human beings are, above all, future-making and at the same time, aswarm with hopes and fears. This observation by a man who was preoccupied with the question of what it means to be human,[2] underscores a *leitmotiv* that runs throughout this book. Namely that of *Homo habilis,* a can-do ancestor of mankind who activates, by his behavior, the inherent potentialities of the universe. However, in understanding this concept, one must avoid succumbing to the notion common in Western traditional philosophy, that the distinctive human capacities for abstraction and system building which we designate by the concept "mind," function merely as a "mirror of Nature" manifesting what is already there, albeit in novel ways.[3]

It is almost as if all we really do is *explicate* rather than truly *create* by our activation and further potentiating these possi-

---

[2] Ortega y Gasset, J. (1987) "International Interdisciplinary Conference" in *Proceedings of the Espectador Universal*, N. de Marval-McNair, ed. New York: Greenwood Press.

[3] Rorty, R. (1979). *Philosophy and the Mirror of Nature*. Princeton: Princeton University Press.

bilities. Nor does being creatively involved in the world mean that we do or can tame Nature to obey our dictates. Creative "orderers" though we be, we are part of an interactive system; as we form, we are, in turn, formed.

Aristotle stated that all philosophy begins in wonder. It would end in wonder, too, if it were not for "Nature's miracle," the creative imagination. Why? because this species specific, i.e., only we humans have it, ability enables us to see not only what is there, but something new everywhere, if only as a possibility! It is a thinking "which builds as it discovers" and will continue to build as long as we are willing to engage in the "loving struggle" it entails.[4] As Einstein observed, imagination is more important than intellect.

Struggle. This is another theme that is woven throughout this book. What is its significance in the context of our concerns? Most importantly, because it is the antidote for our desire to be in "peaceful possession," as theologians were wont to say, of indisputable truths and/or absolute certainties. In reality, this desire is another form of the "will to power" (Nietzsche) and is a far cry from a genuine "will to truth" (Jesus). This latter is an on-going *doing,* a continual self-questioning; in sum, a sense of humility. It eschews the arrogance of power that asserts we can once-and-for-all possess the "knowledge of good and evil" and thus become lords of the universe. To avoid this "original sin" of *hubris,* we need to become comfortable with the discomfort of keeping ourselves "off balance," to exist in the questioning mode. Otherwise, we risk loosing that restlessness which enables us to continue to wonder and to exercise our creative imagination which "displays before us the ever changing picture of the possible" and its relatedness.[5] Thus, the need for thinking in terms of the *Big Picture.*

---

[4] Jaspers, K. (1993) "The Creative Orderers." in *The Great Philosophers,* Vol. III., M. Ermarth and L. H. Ehrlich, eds. New York: Harcourt Brace & Company, p. 188.

[5] Jacob, F. (1982) *The Possible and the Actual.* Seattle: University of Washington Press, p. 67.

Why is this an especially important perspective? Because, unless we embrace it, we risk falling victim to the detrimental effects of *fragmentation*. This is the tendency to take the part for the whole, to enclose ourselves in our own limited, narrow-scoped explanatory system and rest content that we have asked the right questions and given the correct answers, or are at least on the way to that goal. An example is the claim that "knowing the complete human genome we will know what it is to be human."[6] Indeed, we will have reached the state of "eternal rest," of *inertia!* And so we will have reached the end of our struggles.

How tempting. But is it worth it? I hold, along with the author, that the "unstruggling life is not worth living" (*pace*, Socrates). Why? Because it precludes us from thinking in terms of the Big Picture, from *transcending*. Resting content in all-encompassing explanations, we will disengage from the effort to continually broaden our horizons, to embrace as over-arching a view as possible of the interrelatedness of our universe. In a word, we will loose something distinctively human.

Not to be seduced by this temptation demands that we activate our potential for magnanimity, our ability to be great-of-soul enough to undertake the struggle involved. In addition, it demands that we abandon what Paul Zweig has called "the sociology of Narcissus"[7] and embrace willingly the sociology of community, that sense of oneness with each other, both locally and globally, and with our ecosystem. Consequently, it means we must not let our can-do nature result in the exploitation of our accomplishments without any consideration of the long-term consequences of our newly acquired and rapidly expanding technological powers. Essentially, this is a moral struggle and, to achieve this vision, we will be

[6] Schuster, E. (1992) "Determinism and Reductionism: A Greater Threat Because of the Human Genome Project?" in *Gene Mapping: Using Law and Ethics as Guides*. G. Annas and S. Elias, eds. New York: Oxford University Press, p.115 ff.

[7] Zweig, P. (1968) *The Heresy of Self-Love: A Study of Subversive Individualism.* New York: Colophon Books, Chapter 13.

required to develop a much broader, more encompassing moral framework. This will not be an easy task.[8]

The ethical issues raised by modern scientific research, especially in the biological sciences, are unprecedented. And complex as they are, they will become even more complex as this research proceeds. As we struggle with the inevitable ethical dilemmas that will arise, we must keep in mind that the process of solving them is not a simple linear one. As this book makes abundantly clear, we exist in a relationship of reciprocity with the universe. We do not stand apart, nor can we, from this system of mutual input/feedback/output. Therefore, any attempt at devising a new moral framework, must be guided by the nature of this relationship. Given all the factors and, above all, the institutional factions involved, this is a daunting task. We might consider it analogous to a "Copernican Revolution," a breaking apart of the perfectly ordered, hierarchically structured, totally predictable Aristotelian universe. Comforting though that was, it was unreal. The revolution came about because those who initiated it dared to ask different kinds of questions, to pursue different kinds of goals. And just as in the past, we can expect that any attempts to devise a new framework will involve a struggle with the entrenched institutions of moral authority. To work this through with some degree of success, we will, now more than ever, have to rely upon our creative imagination. Only thus can we confront the dilemmas and choices that arise from our creative interventions.

It is just such issues and questions that Dr. Kent's book addresses. I can only conclude by saying *"Tolle et lege,"* pick it up and read it. But beware! Just like St. Augustine, who heeded these words, *Mapping the Human Genome—Reality Morality and Deity* may prod you to undertake the arduous task of undergoing a major conversion.

Francis J. Marcolongo, S.T.L., SS.L., Ph.D.

---

[8] Suzuki, D. and P. Knudtson (1989) Epilogue: "Searching for a New Mythology" in *Genethics: The Clash Between the New Genetics and Human Values.* Cambridge: Harvard University Press.

CHAPTER ONE

THE COMING UPROAR OVER APPLIED GENETICS

Let me begin by expressing my concern that molecular biologists' growing ability to genetically redesign human beings by means of gene replacement will lead to controversies that could reach unprecedented heights of bitterness.

People who are interested in humankind's future need to take another look at what our new role might be as we gradually gain the ability to reconstruct and control nature by altering the genome of all living organisms, especially those of humans. From the viewpoint of evolution, genetic engineering does not differ essentially from the path nature has taken since life began. Nature depends on random mutations of genes to insure the survival of species. Geneticists will soon be able to employ recombinant gene technology purposefully and selectively to achieve the same end.

To some extent the goals of human genetic engineering are rooted in our distant past. Then, as today, people tried to increase their mental capabilities, improve their appearance, and regain their health in imaginative and, often, unlikely ways. Genetic engineering differs in the technology used to attain those ancient human goals. Pre-Columbian Meso-Americans artificially elongated the shape of their skulls to conform to the fashions of their time. The Maori of New Zealand decorated their entire bodies with elaborate tattoos. Up until the early 20th century, aristocratic Chinese bound the feet of female children so that as adults they would be unable to walk and have to be carried by servants, adding to their husband's prestige. It has been reported that a tribe in East Africa fatten young women to make them more desirable for marriage. Today physicians correct congenital malformations, while those who specialize in plastic surgery reconstruct noses, jaws, wrinkled skin, take "tummy tucks," and insert breast implants to help people attain the current ideal of beauty. Future genetic engineering may replace these older methods of altering the human body in response to a seemingly wide-spread desire to change one's appearance.

Let us proceed one step further to creating personality changes. Since ancient times, humans have attempted to transfer characteristics from one person to another. Cannibalism has been practiced in such divergent places as Africa, South America, and the South Pacific islands in attempts to achieve this end. A person's organs were thought to possess attributes such as strength, courage, and cunning. By consuming specific organs, cannibals believed they could transfer their victims' desirable qualities into their own bodies. In ancient Greece, cult members who participated in some of the Dionysian rites consumed human flesh in an attempt to attain a semblance of godliness. The rites were named after an important Greek god, Dionysus, the god of fertility and wine.

Recombinant gene therapy and organ transplants have similar goals. One substitutes healthy functioning body parts for diseased ones while the other replaces defective genes with engineered healthy ones. In the last twenty-five years, we have become accustomed to hearing about successful heart and liver transplants. In time, we will be equally familiar with replacing healthy genes for those that cause diseases.

Eugenics is a theory that was put into practice in Great Britain and the United States in the late 19th Century. Its goal was to improve the human race through selective breeding. Sir Francis Galton, a cousin of Charles Darwin, noted that exceptional ability ran in families and proposed the idea in 1869 in his book *Hereditary Genius*. It advocated the sterilization of those considered unfit because of inherited physical or mental disorders. There is always the possibility that genetic engineering might be used to attempt to achieve similar objectives. A later chapter presents a more detailed discussion of this possibility.

Chapter Six explores personality changes that may also be on the geneticists' agenda sometime in the future. Today, a vast number of people change their personalities, both temporarily and permanently, by means of ingesting drugs or alcohol, or both. If they could escape from some of the conditions of their current life by means of genetic engineering, they might prefer this far more permanent method. Suicide represents a quick, efficient kind of "genetic engineering" that has the added "advantage" of a do-it-yourself capability. It is noteworthy that the number of suicides in the United States continues to rise among adolescents and the aged population. Drugs, alcohol, suicide, and genetic engineering all belong to a class of activities designed to alter aspects of the human mind. Drugs do it by chemistry; suicide does it by various methods used since antiquity; genetic engineering could some day do it by utilizing molecular engineering. It is somewhat startling to think that some of the objectives of genetic engineering have been with us since antiquity.

Let us now examine some of the intriguing facts about the composition of the human genome. Within the nucleus of every one of the approximately 10 trillion cells in our body, we carry approximately 100,000 genes tightly compacted on two winding strands, known as the double helix, composed of deoxyribonucleic acid. (Fortunately, the reader will not have to worry about this term because it has been abbreviated to DNA.) The genes on the human DNA are located in 46 bundles called chromosomes. Bacteria have only three chromosomes but this gives us no reason to feel superior—turkeys have 82. By 1940, scientists had established that the DNA molecule is composed of chemical building blocks in rows of repeated sequences called nucleotides.

The structure of the DNA was first described in the journal, *Nature*, in April 25, 1953, by Francis H. C. Crick and James Watson. In 1962, they shared a Nobel Prize in Medicine for their discovery. Both have written books describing their experiences in the pursuit of the secrets of inherited characteristics. Watson's best selling book, *The Double Helix*, published in 1986, describes how the double strands in the DNA molecule contain the structure of the human DNA. A book by Crick, *What Mad Pursuit*, (1988), looks back on the events in his research.

The invisible DNA molecule would extend close to an incredible five feet if it were stretched out to its full length. The human genome, a term used to describe the sequence of all the genes on the DNA molecule, carries the exact blueprint of the genetic make-up of our species. A person's general appearance such as color of hair and eyes, height, and genetically determined aspects of health, abilities, and personality, lie within the DNA. Biologically, minor variations in the DNA code bestow on us the traits that make each person unique. We also carry within ourselves genes that have no affect on us. We transmit them to our descendants so that a characteristic that did not show up in our lifetime may shape theirs. This happens because organisms have both dominant and recessive genes. The dominant genes give us our own characteristics. But if two people mate who both have the same recessive gene, the characteristic that gene expresses may show up in their children. A single defective gene may sometimes be responsible for a specific genetic illness such as the gene that causes Huntington's disease. Other illnesses and behavioral traits influenced by genes are usually polygenic—that is, affected by a combination of a number of genes each playing a specific role. Often, genes perform their tasks in unexpected and complex ways. A genetic tendency may be activated by a person's environment such as exposure to pollutants, emotional stress, or poor health habits. We must consider gene performance as consisting of a vast number of complex interactions. Both a person's environment as well as genetic tendencies play a role in creating personality traits such as compulsions and depression or performance skills and intelligence. There are specialized regulatory genes that serve as master switches. They send chemical messages to other genes by way of enzymes that cause physical or mental changes to take place. Among these are the genes that turn on the secondary sexual characteristics by triggering the flow of hormones that cause bodily changes during adolescence.

We may never be able to manipulate the genome to produce an Einstein or a Beethoven. Genius, reverence, awe and other higher level human qualities and feelings may, for an indefinite amount of time, remain beyond the geneticist's reach. Nevertheless, geneticists continue to map the uncharted genome and what they may eventually accomplish is presently unknown. Scientists estimate that within the next fifty years molecular biologists will complete the huge task of locating the exact position of every gene in the human genome. Once this has been accomplished, human genes can be catalogued and accessed in the same way as books can in a well-stocked library. Geneticists will then be able to borrow, exchange, and replace genes using the same proteins called restriction enzymes that serve bacteria as chemical scissors. The bacteria use these proteins to cut out of themselves their own genes that are susceptible to antibiotics. Mutated genes unaffected by the antibiotics then replace the susceptible ones. To our consternation these bacteria can become immune to the antibiotic biological weapons we use against them and pass their immunity on to following generations. Molecular biologists can extract the restriction enzymes from bacteria and use them to cut out genes from the DNA of any living thing—ourselves, or any animal, plant, or fungi. The basic mechanism of nature's grand scheme used to pass on genetic characteristics from generation to generation is the same for all of life.

Beyond the requirements of healing, some genetic engineering can reach deeper into the realm of human behavior. A 1993 study in *Science* of more than 1,500 pairs of twins suggests that divorce has now joined the growing list of human behaviors that have been discovered as significantly influenced by genes. Researchers found a variety of inheritable traits that relate to an individual's capacity for happiness, job stability, neuroticism, impulsiveness, and sexual behavior. But some question the validity of this study.

The thought of searching through the human genome for genes that might give their owners criminal tendencies intrigues some and frightens others. If, at some future date, convincing evidence correlates the relationship between specific genes and antisocial tendencies, it may not be long before demands for genetic engineering will be proposed by taxpayers. Instead of building more prisons, geneticists might be able to save the taxpayer millions of dollars and avoid innumerable tragedies if they could replace the genes identified as contributory to criminal recidivism.

A recent issue of the *Psychological Science Agenda*, carried the headline, "Violence Research Produces Controversy." The article stated that the U.S. Public Health Service canceled funding for a conference on genetic factors in crime because of protests by civil rights activists who feared that the results of the research might be misused to increase bias against minorities. Recently emerging disagreements indicate what is ahead. Charles C. Mann writing in *Science* (1994) predicts: "There are few certainties in life. The uproar surrounding attempts to find biological causes for human problems will continue." Under the heading, *War of Words Continues in Violence Research*, Mann reported on the acrimonious controversies over nature's versus nurture's role at the 1994 annual meeting of the American Association for the Advancement of Science. More dissension may occur if an influential scientist or politician proposes that we attempt to use genetics on a large scale to make people content or easier to live with. It reminds one of Aldous Huxley's *Brave New World*, first published in 1932. In Huxley's novel, the Director of the fictitious London Hatchery and Conditioning Center found a method of producing a multitude of identical twins in order to create "social stability." Toward the end of his book Huxley describes them as shouting in unison, "parrot-fashion ...intoxicated by the noise, the unanimity, the sense of rhythmical atonement." The future will decide if Huxley was merely an entertaining author of science fiction or a true prophet of what will come about in the Age of the Genome.

Will there be genetic discrimination in employment? How will employers and insurance companies deal with individuals whose genetic profile indicate the potential for developing mental or physical diseases? Will we have genetic labelling and a genetic underclass? If at some future date genetic engineering becomes commonplace, children may accuse their parents of having failed to give them the characteristics that they wanted. It would then be easy for children to blame their parents for their inadequacies and failures in life. "If you hadn't designed me to be so tall I could have become a space shuttle pilot," a son might complain to his father who had anticipated that his son would like to become a basketball player. "If you had bothered to remove that recessive dumbbell gene from my genome, I could have passed the college entrance exam," his sister might grumble. We are speculating far beyond present capabilities, but it is well to anticipate conceivable problems for which parents might be held responsible in the selection of their offsprings' characteristics.

Several nations have formed watch-dog committees to monitor genetic research conducted within their borders. In our country the National Institute of Health (NIH) and the U.S. Food and Drug Administration (FDA) regulate gene therapy. The U.S. Department of Health and Human Services oversees both agencies to insure safety in the application of human genetics. Other groups also hope to persuade governments to enact laws prohibiting the unrestricted experimentation and manipulation of human genes. However, internationally, genetic research will be as difficult to control and police as nuclear fission. In spite of the requirements for complicated equipment and sophisticated know-how, the possibility exists that illegal laboratories may experiment with the human genome.

With reason, people fear that turmoil, bitter political and religious objections, and, perhaps, even physical conflict may erupt over who has the right to manipulate the human genome and to what extent reshaping human nature should be permitted. The misuse of the emerging god-like power of the geneticist to redesign life may become the dominating topic of controversy in the next century. We can also anticipate competition and dissension in regard to the various methods for manipulating the genetic nature of all life forms. For example, a method of cloning organisms, even people, may some day compete with recombinant gene technology in a bitter struggle to gain preferences and privileges. Controversies will certainly arise regarding whether a given technique is safe to apply or whether it should ever be applied.

We are entering an age where, in theory at least, we shall be able to genetically redesign ourselves. The temptation to do so will be difficult to resist. Sailing uncharted seas, climbing up to mountain tops that were thought to be beyond human reach, penetrating the Arctic wastes, placing a footstep on the moon challenges us. Can we do it? Dare we? Should we try? These questions have always been answered in the affirmative at some time in our history. Why would an attempt to reach paradise by way of gene technology be an exception? In the next chapters we shall examine reality, morality, and deity. It may be prudent to do so before we attempt to use gene technology to regain paradise by reversing the consequences of the temptation to which we are told Adam and Eve succumbed. On the other hand, genetic engineering might represent a modern forbidden fruit that will deprive us of whatever remains of Eden in our present world. We begin to look for answers to these questions by exploring in a new light the age-old riddle—what really is real?

CHAPTER TWO

THE CONUNDRUM OF REALITY

Even before the dawn of civilization, humans probably wondered about a spirit world that seemed neither real nor unreal. This chapter gives a brief overview of the history of thoughts on reality and explores the directions into which various perceptions of reality may lead us in the next century. Reality might have been questioned long ago when humans scanned the horizon and saw a distant lake that upon investigation turned out to be a mirage. Perhaps reality perplexed the Neolithic man who dreamed vividly one night after a hunt, that he had returned to his cave with bison meat—but upon waking the next morning, could not find it.

Much later, when humans became sedentary and could afford to support full-time philosophers, they became aware that the nature of reality was elusive and argued about what reality really meant. Throughout the ages debates on what reality consisted of continued. Contrary to what philosophers would have expected, particle physicists, in the early part of this century, found it necessary to turn to philosophy to account for the unexpected behavior of subatomic particles. Physicists noticed that the tiniest bits of matter were always seen spinning into the direction opposite to the angle from which they were observed regardless of the position taken by the observer. It didn't make good scientific sense for particles to respond to the way human beings looked at them. Particle physicists concluded that something in the human mind had to be involved in the scenario.

In 1927 the German physicist, Werner Heisenberg, proposed his astonishing Uncertainty Principle. It entailed the measurement of a quantum system, such as an electron or an atom. It held that any measurement of such a system would disturb it, causing it to become unpredictable. The amount of disturbance was found to be proportional to the precision of the measurement so that the more precise the measurement, the greater the disturbance. The implication was that the laws of atomic physics formulated probabilities instead of certainties as had previously been assumed. After that, physicists could do nothing but conclude that reality was, in part, created subjectively. That notion, of course, put them knee deep into philosophy. However, I must mention that Gary Taubes, entitled a two-page article in *Science*, (1994) *Heisenberg's Heirs Exploit Loopholes in His Law*. Probing for reality whether in the sciences or in philosophy seems to be a never-ending job. Few concepts have been given so much attention or searched so systematically for hidden meaning by philosophers as the nature of reality.

On the other hand, many down-to-earth people prefer not to think about reality at all. They believe they can live very pleasant lives without worrying about it and can easily prove they are right. Yet reality does matter in a way not immediately apparent. In fact, reality influences everything people do and that includes even those who are too busy to be concerned about abstractions. Unknowingly we all live our lives according to our perceptions of reality even as we show our children how to kick a football or kiss our mate goodbye when leaving for work. We avoid thinking about or doing anything about things we do not accept as real and this may profoundly influence the way we live our lives.

But meaning is grist for the mills of philosophers. The word *philosopher* is taken from the Greek and means "lover of wisdom." There can be no sensible discussion of meaning among any group of people unless it is accompanied by an assumed consensus on what constitutes reality. Nevertheless, some philosophers skim over the definition of reality as they expound on less recondite subjects. This causes dissention among them they may have avoided had they first reached an agreement on the nature of reality. Some philosophers simply hang the label *reality* on whatever happens to be in their line of observation. Since ancient times, lovers of wisdom have faced different directions and this has left room for argument. In the 5th century B.C. Plato, one of the best-known philosophers of all times, may have startled his students by claiming that genuine reality couldn't be found in anything that *looked* real. Instead, reality had its own independent existence only in forms or ideas that he called, archetypes of visible things. Things that could be observed by people were merely "shadows" of abstract essences. It wasn't long before one of his famous students, Aristotle, disputed this, asserting that all things people observed were real. He observed that if you stepped on a cat's tail it didn't react at all like a shadow.

The wise men of the East view such musings as delusions. The Hindu spiritual leaders teach their followers that Ultimate Reality is beyond human understanding. It defies description and can't be seen in the everyday world because it isn't accessible to human senses. The Hindus consider Ultimate Reality, called Brahman, vast and remote. One reaches it only through liberation from the cycle of reincarnations by living an ascending order of increasingly blameless lives.

Western scholars have championed their own fragments of reality throughout the ages and given them a variety of names. Depending on their particular focus, they called reality: objective, subjective, empirical, abstract, concrete, physical, sensory, relative, quantum, virtual, primordial, religious, existential, and so on. Although these scholars referred to their fragments as "reality," they represented only an assortment of categories and not reality in its larger sense. If a person has a foot planted on one fragment of reality while his other foot rests on a different fragment, he will lose his balance when these fragments drift apart, as they usually do in time. Should he look heavenward instead of down at his feet, he will be forced to invent an innovative philosophy to account for his discomfort. That situation should be included in textbooks as a warning to future students of human genetic engineering bent on improving human nature. Albert Einstein once claimed even scholars of audacious spirit and fine instinct can be obstructed in the interpretation of facts by philosophical prejudices, stemming, I will add, from the fragmentation of reality.

How can we come to grips with reality in everyday life experience is a fair question. Those who search for reality in exotic places resemble people who look for their misplaced glasses while wearing them on their noses. We may fail to recognize reality because it is all around us. It makes some people uncomfortable to think too much about the nature of reality just as many of us get tired of too much discussion on the nature of truth. However, when thinking about the consequences of impending genetic engineering we must be willing to view reality in its largest context. This means a Big Picture view that includes all the brush marks composing it.

Our concept of reality must reach out beyond self-involvement and search for its outer limits within the broadest possible framework. For example, confining our definition of air to that which escapes from a punctured tire ignores the fact that only a small amount of all the air in the world was pumped into the tire. Christopher Frye, a 20th century English dramatist, wrote in his play, *A Yard of Sun*, (1979) "There may always be another reality to make fiction of the truth we have arrived at." That fiction represents only a fragment of reality which distorts it. This applies to all partial views of reality. If we use an analogy in which air represents reality, it should include more than the air in our own tire but all the air in all the tires of the whole world as well as to the air not in the tires.

I believe that a unity of purpose in applying gene technology must be achieved against a background of an unfragmented reality. Such a reality is all-inclusive and consists of both fiction and fact, physical and metaphysical. It consists of Plato's archetypes together with all of their "unreal" shadows. It includes every act, thought, dream, scrap of imagination, hallucination, stick, stone, worm, rat, cat, and every human being ever born. All things that happen and have happened—all happenings, as I like to call them—are building blocks of which the universe is constructed and constantly reconstructed. Happenings make the universe what it was, is, and will be. Every event, be it a leaf dropping or a star collapsing, shares a common heritage with all other things that have occurred. Everything that came into being emerged from the same vast, immeasurable pool of unborn potentialities. In becoming real they are exclusive since there is, also, non-potentiality which can never give birth to reality. In chapter 5, I shall discuss the important role of nonpotentiality in human lives. Its significance in genetic engineering is important since it colors our perception of the universe itself. Beyond strictly human concerns, potentialities will play a decisive role in how we will relate gene technology to all of nature.

Most of us have the false mental picture of the past as disappeared and gone. The English poet, John Keats, wrote: "A thing of beauty is a joy for ever...it will never pass into nothingness." Continuing in a poetic vein, I agree that all happenings, beautiful or not, leave footprints in the sands of time that cannot be erased. Each unborn potentiality of the universe that is actualized into reality contributes to the ever expanding content of the universe. If billions of years from today, the universe collapses into itself and shrivels into a black hole consisting of a featureless radiation that wipes out all past information, or rejuvenates itself to maintain a steady state, the reality of irretrievable past events cannot be extinguished. All things change, build up, and wear down; are transformed, transcended, or metamorphosed. Irreducible energy, if used, yields to entropy. The present yields to the future and becomes the past. Yet neither energy nor reality can be eradicated.

Each step in the process of change continues to live as a participant in the immense parade of events representing actualized potentialities. The reason for this is that reality incorporates its past within itself and retains it. In preparing the mind for the new age of the human genome, we must become receptive to the idea that heritage is not confined to the history of our family or our nationality or our socio-economic status. Our defective genes, even if replaced, will remain as things in themselves and wonders of nature whether they were good or bad for us.

This way of thinking about reality could provide us with an insightful outlook. Let us imagine that a person writes a poem and submits it to judges of a poetry writing contest in hopes of winning an award or getting it published in an anthology of contemporary poetry. Let us assume that the writer forgot to make a copy of his poem and that it never reached its destination. The distraught poet may think that he accomplished nothing by his effort. However, his hope of winning a prize, the challenge, the ideas and feelings expressed in the poem, even the expected admiration of his friends, remain within the reality of his anticipation. Living in our Western culture conditioned him to view winning as one side of the coin of achievement, and worthlessness as the other. However, the feelings he experienced continue to live, not only in his memory, but in the reality of having existed. If one could accept it, this perception could become important to a person, not as denial nor as a rationalization, but in a positive way as an alternate view of "failure." Then Gray was mistaken in writing in his famous *Elegy Written in a Country Churchyard*, when he wrote, "the paths of glory lead but to the grave." Instead, the "paths of glory" remains for all time a path along which the universe traveled in becoming what it will eternally consist of.

The fear of death has often been ascribed to a fear of the unknown when actually it represents an anxiety stemming from the seeming loss of opportunity for continuing to actualize the potentialities of the universe. That is why people who feel they have used their opportunities to actualize potentialities of the universe believe, as they might say, they made a difference in their lives and are able to accept death more easily than others. Lao-tzu wrote in his book, *Chung-tzu*, in 640 B.C., "A man is ready for sleep after a good day's work." This view permits us to feel that we can gain satisfactions without fame or recognition just by virtue of the fact that what we did *occurred*. The death of a young person who did not yet have the opportunity for self-actualization strikes us as a greater tragedy than the death of an older person who had the opportunity to actualize potentialities in life.

The influential 16th century French philosopher, Descartes, may not have agreed with me. He did not accept any happenings as reality without proof and even questioned the existence of any reality at all. Like many others he was convinced that events had to be demonstrated tangibly to become real. This view of reality is the extreme opposite of the one I have described. Finally, Descartes felt compelled to search for evidence of his own reality. Reluctantly, it seemed, he had to conclude that his ability to think proved that he existed.

Some present-day philosophers believe that by turning to the neuroanatomy of the brain they can locate the origins of philosophical truths. They try to find the sources of human consciousness within neurons, synapses, and neural networks. But they will not find consciousness locked away within anatomy. Physiology relates to consciousness in the same way as existing does to thinking. Descartes would have said, "I am conscious; therefore, I am alive," when he should have said, "I am conscious and I am also alive."

Reality is sometimes confused with truth. But in my view. both truth and lies create reality. That is why lies, in the minds of some, easily convert to "truths" that they will defend vigorously even with their lives. Some can pass lie-detector tests without difficulty because although they know their statements are fictitious or half-truths, to them these assume the emotional characteristics of truth. Thus some people experience no guilt when they lie.

Reality is not created by action alone. Inaction creates reality as well. If a man digs a hole using a shovel, he creates reality by what he does. If another man with shovel in hand simply stares at the ground, he creates reality by what he doesn't do. In each case something that could occur did occur, and occurrences make reality. An unusual creature called a Tardigrade can exist for more than a hundred years without water, oxygen, or heat in a state which by most definitions would be called "death." Yet when moistened it immediately springs back to life. Some 1 billion-year-old fossils of blue-green bacteria and their modern counterparts which form a living film on stagnant water look exactly alike. The survival of this species without apparent changes over such a length of time actualizes a potentiality of the universe just as do new species that undergo rapid changes.

In psychiatry schizophrenic patients are described as having lost contact with reality. The description serves to communicate the state of the patient's mind but taken literally reflects the prevailing misconception of reality. One cannot lose contact with reality. Instead, the patients' hallucinations and delusions represent realities to which their doctors do not subscribe.

To illustrate I offer an analogy that certainly seems ridiculous. In a hypothetical scenario one can merely imagine in one's mind one is eating a meal and consider it as real as actually putting food into one's mouth and swallowing it. Certainly merely imagining that one is eating a meal would be of no satisfaction to a hungry person. Regretfully, in this world, satisfaction and reality often remain independent of each other. No one would feel less hungry if told that fantasizing eating a meal activates more cortical neurons in higher brain centers than actual eating does. Nevertheless both of these alternatives create reality even if one of them eventually leads to starvation.

Surely, a difference exists between an actual stone picked up from the ground and one that a person only imagines he is holding in his empty hand. The only difference consists of neural messages from the person's brain that communicate "stone" for only the actual stone and not the imagined one. In matters more important than stones, we use a different method of ascertaining reality. A psychologist utilizing established norms may objectively evaluate a person. When the person receives the results of his evaluation, without psychological intervention, he is likely to continue to respond to his own self-image—the imagined stone, not to the "real" stone in his hand. The sugar-filled pills we call placebos have worked remarkably well in reducing many symptoms and also have created side effects that only genuine medications are supposed to cause. Computer stimulated "reality" creates visceral and sensorial perceptions of reality that bypass skeptical and rational faculties. Even those who claim that only objective reality is real will have difficulty identifying it.

Let's look at subjective reality now. In T. S. Eliot's *Murder in the Cathedral*, (1935), the Four Tempters who personify human inclinations lament, "Man's life is a cheat and a disappointment. All things are unreal. All things become less real, man passes from unreality to unreality." The tempters suffered from a common misconception. Their "unreality" represents the most poignant aspect of reality. Hopefully, future neurogeneticists will respect the idea that our most human qualities come from that element of the "unreal" that represents the greater-than-the-sum of our anatomy. Genetic engineering has been likened to a "magic bullet." But human behavior more closely resembles the movements of a Ping-Pong ball batted back and forth in a game played by amateurs than to the trajectory of a bullet fired at a target by an expert marksman.

In making momentous decisions, considerations must be given to the reality of things as they are *not*, as well as to things as they are. Human behavior is rooted in imagination, daydreams, and fantasies as well as in events that are concrete and objective. The here-and-the-now mind will never grasp the significance of the important abstractions that play a part in the way human genetic engineering will be conducted. Our own modern culture undervalues daydreams. A society that values deeds too highly and thoughts too little produces people with superficial satisfactions and underlying discontentments. Could this explain Thoreau's often quoted observation: "The mass of men lead lives of quiet desperation." Now is the time to ask, will someone in the future try to alter with gene replacement the desperation that our culture—not our genes—has given us?

I believe we cannot know the universe in its totality because it represents an abstraction that our cognitive capacity cannot grasp. If we reduce its essence to the level of our subjective perception it allows us only partial views. However, there is something akin to the beyondism of Ultimate Reality within us. It is precious because it enables us to identify with the universe—experience it—in spite of our limitations. Genetic engineering represents a biological technique and, in that respect, it is a one-dimensional approach to the multi-dimensional human beings it intends to serve. In the long run, only an unfragmented view of reality that includes our feelings as well as our knowledge will enable genetic technology to serve humankind. Knowledge and techniques represent fragments of reality as do all things viewed as isolated and unconnected. The danger exists that in planning to modify the human genome, fragments of reality may masquerade as the whole of reality. Should the pretensions of any of these fragments beguile those empowered to apply recombinant gene technology to humans, they will fail to see that reality extends beyond the coded messages on our DNA.

This chapter suggests that if we take the Big Picture view of life and the universe our acts and thoughts must be seen as contributing to the reality that creates the history of the universe. It is a sobering thought that we are in part responsible for that history by how we live our lives. In the next chapter we shall look for the characteristics in human nature that prompt us to seek challenges. It is appropriate to do so at a time when the judicious application of gene technology offers us one of the greatest challenges we have had since we evolved as a species.

CHAPTER THREE

A FATEFUL URGE TO REDESIGN OURSELVES

Genetic engineering has enabled us to design insect-resistant plants that improve the quality and quantity of our food supply. Some think there may be a hidden danger in eating these foods because altering them genetically may cause unpredictable side effects. Others have religious objections to "meddling" with nature because they view genetic engineering as an affront to a Creator's handiwork. They believe that changing life forms is not a human prerogative.

However, as gene technology increases the world's food supply, these objections will be overridden by pragmatism and the successful results of genetic engineering. I do not expect a universal outcry against genetically modifying the bugs, snails, and worms that harbor viruses and bacteria dangerous to humans. Current researchers are working on a method of genetically altering mosquitoes so that they will no longer be able to transmit malaria. An expanding zoo of transgenic animals: mice, pigs, and bacteria—all-star performers in genetic engineering—now manufacture human enzymes, proteins, and replacement genes in their bodies. The animals have been involuntarily enlisted to help combat an estimated 3,000 human genetic diseases. Genetically altering animals to reproduce human genes may seem somewhat distasteful but, so far, this use of gene technology has not been seriously opposed.

As plans for genetic engineering become increasingly ambitious, it will become correspondingly more difficult to avoid objections to its goals. A philosophy of genetics now knocks on the doors of university departments of psychology and sociology asking to be admitted to disciplines that traditionally championed the importance of environment in shaping behavior. Meanwhile, the arguments on the relative influence of environment versus heredity—the old nature/nurture controversy—continues unabated.

With the continued advance of genetic engineering, the controversy takes on a new urgency. Genes may play a role in some diseases that were previously thought to result solely from exposure to poor environmental conditions and imprudent life styles. New evidence suggests that certain genes lower resistance to heart disease and cancer. In time, these mainline killers of people may join the growing number of human diseases that can be prevented or even cured by injecting patients with viruses carrying healthy copies of their defective genes.

Many social and ethical problems will accompany the expanded application of gene technology. Recently, the media reported a case of a 51-year old woman diagnosed as having a terminal brain cancer. Her doctor wanted to administer an experimental gene therapy previously untested on humans. When the National Institute of Health received the request, a debate at the Recombinant DNA Advisory Committee brought to light numerous misgivings among its members. Concern was voiced that approval might open floodgates to hundreds of similar requests from doctors of dying patients seeking permission to try unproven genetic therapies.

We can imagine the debates that might accompany the use of gene technology if research could prove that this method could cure disorders that affect the human mind. Some scientists suggest that there is evidence that inherited susceptibilities underlie the development of some psychiatric diseases such as manic depressions and schizophrenia.

The association between a specific genetic defect and a personality disorder was shown in a study of 104 adults and children from 18 families reported in *The New England Journal of Medicine* in 1993. Seventy percent of those with a specific defective gene were found to have attention-deficit/hyperactivity disorder. Children with this disorder tend to be inattentive, impulsive, and restless. They experience difficulty in school and social situations because they lack the ability to focus sufficient attention on what is going on around them. The genetic codes for a defective thyroid hormone receptor were identified as responsible. In time it is likely that many more behavioral traits will be linked to a specific gene or group of genes.

For many years childhood conditioning and aggressive role models were held responsible for later, antisocial adult behavior. Violent crimes were viewed as caused solely by sociological and psychological factors. Now some research has linked acts of aggression to individual genetic makeup.

When scientists sometime in the future attempt to develop gene replacement therapies to modify or prevent undesirable aspects of human behavior, they will probably meet with resistance because some will fear that genetic emphasis will minimize important environmental factors that also contribute to shaping behavior.

The use of gene technology will be spurred on by an irrepressible human urge to explore the unknown and discover how far we can push the outer limits of human achievement. Scientists as well as many of the rest of us may succumb to the gambler's obsession to beat the odds. Among those who have done so are the majority of Nobel Prize winners. Breaking records fascinates modern humankind even in trivial matters. Some were bent on discovering how long a member of our species could perch on top of a flagpole—a craze popular just a few decades ago. These quests involve more momentous aspects. As we establish new records of human endurance, wisdom, and even folly, often we do not realize that they are rooted in the dynamics of the universe—and that everything humans do or think actualizes additional potentialities that shape the universe's changing configuration. After our ancestors developed agriculture about 12,000 years ago, they seemed determined to discover what other spectacular feats they could perform. Excitement accompanies discoveries that actualize new human potentialities and contribute to our technological breakthroughs from digital telephones to heart transplants. The Egyptian and Mayan builders of pyramids and the people of Britain who constructed the Stonehenge must have experienced this same excitement. Today we often succumb to the lure of a challenge and overcome our fears of the risks involved, some even willing to die in order to accomplish the "apparently" impossible. The Age of the Genome will offer us an unprecedented opportunity to do this.

In the absence of natural challenges, people create artificial ones. We engage in competitive sports, train to beat all kinds of records, test our individual skills with games, or puzzles, and if nothing else, argue fiercely over trivia. Sometimes people champion causes as much for the sake of a challenge as because of conviction. Curiosity is a motivating ingredient. Curiosity deflects boredom for higher forms of life but more important, motivates them to actualize their potentialities. Curiosity becomes an irresistible force when it stimulates the approximately nine billion synapses in the human cerebral cortex. However, if curiosity were to play the major role in genetic experimentation, it could lead to unpredictable and, at times, less than desirable consequences. Curiosity and humanitarian motives have different roles in human nature, but in the pursuit of science they are usually felicitously combined. Yielding too much to curiosity in genetic research, however, may inadvertently lead to changes that would not benefit humankind.

Some environmental changes in climate and a lack of ecological niches have led to adaptations that created a proliferation of new species and variations in forms of life. To insure survival, a tiny flea may lay 400 eggs; a single fish lays thousands. One teaspoon can hold billions of viruses. Nature has two kinds of responses to evolutionary pressure. One is adaptation through change in behavior when circumstances require it. The other is of conserving the *status quo* when the circumstances permit it. One contributes to diversification, the other to maintenance of what has been achieved. There have always been "stayers" as well as "movers" responding to evolutionary pressure. Some scorpions living today have changed little from their 380-million-year-old fossilized ancestors. Likewise, among us there are liberals as well as conservatives and we can predict that their agendas for genetic engineering will certainly differ.

Forty years ago, Lecomte du Noüy, a biologist and philosopher, wrote a book titled, *Human Destiny*, (1949). Many people today would agree with what he wrote. Du Noüy complained that the material side of civilization has conquered "the soul of modern humanity" and keeps people "in a kind of breathless expectation of the next day's manufactured miracle." As a result he found that, "Little time was left for solving the true problem: the human problem." By "human problem" he meant the difficulties we have in getting along with one another, the rivalries between nations, their wars, their poverty, and our competitive acquisitiveness. Like other critics of our Western world's lifestyles he was unimpressed by the idea that civilization's gadgets and inventions may help solve the very human problem, boredom, to which I have referred earlier. Each "miracle," electronic or mechanical invention actualizes a potentiality that exists somewhere in the universe and discovering these play a role in its appeal in addition to its utility. People often buy things that are clever even though their utility is minimal. There is always something thrilling in the thought: Wow! This thing works! It did not occur to du Noüy that some material and spiritual things might even be extensions of each other. Human ability to invent gadgets is sometimes awe-inspiring as archaeologists will probably think a thousand years from now.

Some years ago a motion picture, *The Gods Must Be Crazy*, depicted a Bushman (a more correct name is Basarwa) living in the Kalahari desert in Africa who found a discarded Coca-Cola bottle. He thought that the gods must be crazy to leave such a magnificent object behind. The Coca-Cola bottle story was merely a Hollywood scenario. Nevertheless, the story makes the point that we fail to appreciate our abilities. There has been so much internal and external criticism of our materialistic attitudes that we have lost faith in ourselves. It isn't surprising that we don't appreciate the achievements brought about by our modern technology.

Some criticism of our technological life style is undeserved and stems from a lack of comparison with alternative ways of living. The eloquent critic of Western technology, du Noüy, (1949), likened our enthusiasm for technological innovations to the fascination of children "in their first view of a three-ring circus." To be sure, a carnival-like media atmosphere seems to accompany breakthroughs achieved by molecular biologists. It is justified that we celebrate steps that bring us closer to understanding of our genetic selves. Lecomte du Noüy wouldn't agree. Were he writing today he would probably predict that future "three-ring circuses" would feature performing humans who had been genetically engineered to jump through hoops. They might even replace lions and tigers currently trained to do so. In following chapters we will explore to what extent such fears are justified.

Socrates' often quoted dictum, "the unexamined life isn't worth living," leaves a key question unanswered. What kind of an examination is required to make life worth living? What textbook should be selected to prepare one for such an examination? Should we read books on religion, philosophy, Aesop's fables, the *Encyclopedia Britannica*, or must we read them all? When 21st century geneticists gain the ability to redesign our minds and bodies, the steps they will take—if any—will depend on the "textbook" they will use. And that, in turn, will depend on what school of thought will have influenced them. Modern medicine is one school of thought. Modern medicine's venture in providing life-saving skills and medications to developing countries illustrates the ironies that might be duplicated in the Age of the Genome. Medical efforts in these countries deserve recognition for saving an unprecedented number of children's lives. But the resulting population explosion partly contributes to the problem of widespread hunger. I do not blame modern medicine for the fact that half of the world's children are malnourished and many face ultimate starvation. However, it is well to remember that every actualization of a potential serves as a catalyst for actualizing additional, often unanticipated, potentialities. Genetic engineering may trigger a series of unintended, unexpected, and unpredictable events that result in chaos instead of order and progress.

Chaos represents a relatively new concept in science and has become a field of study in itself. It differs from the earlier idea of cause and effect in that it utilizes nonlinear mathematics which seems to reach conclusions by a circular method. This kind of mathematics can deal more effectively with events when the final outcome cannot be predicted by traditional, straight-line mathematics to which we are accustomed. Chaos has been defined as total disorder and as infinite, formless space. More recently it has taken on the meaning of the unpredictability inherent in complex interactions of nature. The notion of chaos suggests that a simple event can have complex, unpredictable results as each event triggers subsequent unanticipated events. Chaos is present in cosmology, chemistry, meteorology, epidemiology, and has more recently been found in psychiatry and psychology. We are faced with the possibility that every application of genetic engineering may have unanticipated chaotic consequences, aftershocks, and reverberations before the intended goal can be reached. Each of these steps represents a new actualization of a potentiality that creates our eternal reality.

Anthropologists Louis Leakey and his wife, Mary, discovered one of our 1.6 million-year-old-tool-making ancestors,

Homo habilis, at Oulduvai Gorge in Tanzania. Mary Leaky named the fossil *Homo habilis* which means the "the handyman" or, loosely translated, the human who can do. Much older human-like fossils have been discovered since then including the *Australopithecus afarensis,* viewed as a subgenus of *Homo sapiens* reported by Shreeve in *Science,* (1994), as "arguably the oldest human ancestor of them all." The limb bones were suited for climbing in the trees and also for walking on the ground on two legs. A diminutive 3.2 year old primate, popularly know as "Lucy," found at the Hadar site in Ethiopia, lived approximately 3.2 million years ago and has, by some human paleoanthropologists, been viewed as "the missing link." The dawn of humankind will always remain controversial as new finds are discovered and previous ones re-evaluated. Throughout the rest of the book I shall refer to *Homo habilis* not as a specific fossil but as a generic term for our earliest toolmakers whoever they may have been. Our ancient ancestors made only crude hand axes by striking stones with other stones. Nevertheless, this was a momentous beginning. Apes in the wild modify objects as twigs and stones to use as tools. But only humans use tools to make other tools. The idea caught on as over time our brain capacity increased by more than seventy percent from the earliest fossils believed to have belonged within the human line. With their enlarging brain humans became gifted, can-do actualizers of those potentialities of the universe that without them may have remained dormant forever.

The idea of doing creative work on earth has been put into a religious context since ancient times. When Bishop Tutu of South Africa was asked to define the meaning of life he replied, "It can be found in manifesting the glory of God." Actualizing the potentiality of the universe is a way of putting this same thought into a non-theological context. A San Diego church bulletin recently quoted an unknown poet: "Wondrous things lie deep within the world and are waiting for us to uncover." What are these wondrous things? Gadgets? Ultimate Reality? Mystical experiences? A mapped human genome? The answer is—all of these. Later chapters will show that immensely more profound potentialities in the realm of spiritual values are within human grasp, but as we shall see—precariously so.

Biologists use the term "evolutionary pressure" to explain why plants and animals redesign themselves over time through the mutations that have resulted in evolution. Evolutionary pressure may also be thought of as the push that forces us do the incredibly arduous and time-consuming job of mapping the human genome. Time will tell if our ability to use gene technology might enable us to flee the earth and escape to another planet if astronomers were to warn us that a huge asteroid on a collision course was headed our way.

The threat of extinction works like an invisible hand that prods all living creatures to actualize an increasing number of potentialities of the universe. The result, according to the 19th century English biologist and philosopher, Herbert Spencer, insures "the survival of the fittest." This negative view of the survival of life leaves a false impression. Evolution depends instead upon mutations—life savers as well as killers—that have sometimes been called "nature's mistakes" because only a few out of many have proven to be beneficial.

But mutations are not always mistakes! Over time some have been brilliantly successful in preventing the extinction of their host organisms even though haphazard, hit-or-miss gene alterations accomplished this. Mutations, in the "bugs," as doctors call our diminutive enemies: viruses, bacteria, and fungi —have found ways to combat our most ingenious efforts to eradicate them. Medical geneticists continue to search for new genetic weapons to fight them, as they search for ways to correct the damage caused by our own mutations. Genetic research may further our own chances for survival or, conversely, satisfy a curiosity that might override prudent caution. This book was written to contribute to other efforts aimed at avoiding this possibility.

In summary, the dynamism of the universe is reflected by the forces operating within it. Life has evolved because a few random, favorable mutations among many unfavorable ones created adaptive biological changes in primitive living organisms. However, new studies raise the question of some limited mutational selectivity in the process when organisms are under strong evolutionary pressure, *Science*, (1994, 265: 318-19). In contrast, genetic engineering is both mind-controlled and goal-directed. But history has shown that human minds can be fallible and goals are not always attainable.

Biologists have described nature as "innovative." Innovations actualize the latent potentialities of the universe occurring within the flow of events. This aspect of nature has captured the human imagination. Cultures, legends, and oral traditions of nonliterate peoples as well as the can-do optimism of Western technology reflect the spirit of human innovation. We feel the need to explore the unknown world, climb the highest mountains, pursue subatomic particles and map the human genome. The urge to redesign human beings and to place a human footstep on the moon have much in common. In that respect, we can be reasonably sure, our archetype, *Homo habilis*, would approve of us. Of course, he couldn't have known that his can-do beginnings would eventually lead us to map our human genome and give us the capability to redesign our species. Had such an idea been able to penetrate his comparatively thick skull, we might not be at all certain that we could count on his approval.

CHAPTER FOUR

THE EGG AND THE ROOSTER'S CROWING

Along with the can-do optimism and accomplishments that accompany our technological breakthroughs, our Western civilization must also cope with disturbing and seemingly insoluble problems in our daily lives. We should not be surprised that opposition to the influences of Western civilization surfaces in many parts of the world. People vainly search for a meaning in life beyond success in their employment and professions. Reasonable decisions on how to apply genetic engineering to human needs are difficult to make in a world of dissension and disillusionment. The present incongruous mixture of optimism and discontent cannot avoid causing bitter conflicts over the use of gene technology in the Age of the Genome.

Today there is hardly a newspaper editorial page that fails to tell us what is wrong with our society. Religious leaders of many faiths call our attention to what they identify as a contemporary spiritual and moral decline. Sociologists point out that the mobility of our society deprives people of the traditional guidance and support of their immediate and extended families. Signs of our contemporary malaise include the rapid increase of violence, the widened generation gap, and the destructive drug culture.

Multiple factors contribute to our sociological problems. Usually they are complex and subject to a number of interpretations. Most people would agree that the flow of past events in the history of Western civilization has contributed to many of the problems we face today. Explanations for today's difficulties using a time frame of one or two generations do not permit us to view them within the larger framework of their chronological development. An awareness of their historical background will not solve them but may provide the perspectives we need to gain a deeper understanding of how they came about. This could lessen the danger of passing on distortions and oversimplifications to the generations who will live in the Age of the Genome.

Setting historical time frames for sociological events has to be arbitrary. Rarely in human history can we point to a specific event and say, "Here's where the problem began." Nevertheless, let us look back to the period in Western civilization known as the Middle Ages that followed the collapse of the West Roman Empire in the 4th and 5th centuries. This period was not as homogeneous as its name implies—the "Dark Ages." There were times of great intellectual ferment and innovation in the arts and literature in the late Middle Ages.

The dominant system of social structure consisted of three classes—nobility, clergy, and serfs. Land was owned by kings and nobles and worked by serfs. Sometimes land was granted by overlords to vassals who swore allegiance to them in ceremonies that cemented their personal relationship. The feudal system was essentially a military organization with knights as warriors. Christian and military ideals created the morality of the times. Chivalry was an ethical ideal that required the knights to take vows of fealty to their lord and, in theory, protect the weak and poor. According to tradition, one of the objectives of knighthood was to impress women of their class with their exploits of skill and bravery. This was often accomplished on horseback in tournaments where knights knocked other knights off their mounts by crashing into them with poised lances. Throughout history the human can-do spirit finds outlets that seem strange to people living at a different time.

The intellectual forces in Western Christendom that held medieval civilization together was called scholasticism. It consisted of theological teachings mixed with Greek philosophy. This was a time of the exploitation of agricultural workers, religious dogma, and superstition. There were ceaseless raids carried out by nobles who obtained their land and power largely in battles with one another. Nevertheless, feudalism offered the people, including agricultural workers, a personal identification that is missing in our contemporary culture. The paternalism of lords and clergy gave serfs a sense of connectedness within their own microcosm of society. Even as serfs were exploited, they found security in knowing who they were and what they should believe. As commerce increased in the late Middle Ages, people were drawn into towns and cities where merchants gained political power and formed a fourth social class.

After merchants had gained increased social status, an intellectual revolution occurred in Europe in the 15th and 16th centuries. The revolution began in Italy but became known by the French word for rebirth, "Renaissance." Independent city-states increased their political and economic power during the prevailing period of relative social stability. Urban centers of population established wide contacts with other cities and states. The revolution was hailed as a liberation for the people of Europe. It was a time of brilliant accomplishments in scholarship, literature, and the arts. It provided people with a new world-view of human destiny. The intellectuals living at the time of the Renaissance would not have believed that anything but good could result from their newly appropriated sense of freedom. Ironically, one of the most benevolent periods in European history noted for the liberation of the human spirit may have helped shape the distinctive characteristics of our contemporary social problems.

In 18th century Europe the Age of Reason followed the Renaissance. Intellectuals attacked religious intolerance and censorship. After hundreds of years of hearing themselves referred to as sinners, people were told that they themselves had the capability to make the right decisions about their lives. A contagious optimism swept through Europe during these years known as the Enlightenment. People believed that if scientists had sufficient information, their rational minds could solve all human problems.

Innovative ideas made it an exciting time for intellectuals. The people of the Enlightenment no longer believed that they lived in an earth around which all planets, stars, and galaxies revolved. The English mathematician, Sir Isaac Newton, born in the same year that Galileo died, discovered the laws of universal gravitation. His many accomplishments included the formulation of his three laws of motion which helped make physicists aware that order exists in the universe. When evidence and proof replaced intuition and false certainty, science flourished.

The new doubt about the nature of reality elated some people but disturbed others. Descartes, the philosopher I referred to in chapter 2, lived at about the same time as Galileo (1564-1642). Descartes believed that nothing should be taken for granted and that even human existence must be doubted. Later, the influential Irish philosopher and clergyman, Bishop Berkeley (1685-1753) proposed that there was no existence beyond perception. He wrote, *"esse est percipi"* (to be is to be perceived). He maintained that perception was the only source of reality and that only seeing justified believing. A tree isn't a tree until someone observes it. Some people were convinced that Bishop Berkeley was right. However, the idea that happenings become real only by virtue of someone's awareness of them breaks reality into fragments. In time, the fragments of reality that visibility represented took the place of total reality in many people's minds. This distortion of reality continued and still exists today. It helped to create the philosophical underpinnings of our contemporary problems.

The skepticism that accompanied the new freedom of thought caused many people to turn from faith to science for an explanation of life's meaning. In time, people decided that only doubt was capable of unearthing basic truths. The influential British philosopher, David Hume (1711-1776), insisted that most of the principles of our knowledge had no justifiable basis. He believed that before any idea could be accepted, proof available to the human senses had to exist. Many people living in that period, as some still do today, ignored the fact that science too represents only a fragment of reality. The new doubt about the nature of reality elated some people but disturbed others. In their confusion, people no longer knew their roles in life as they had previously under feudalism. They felt compelled to search for evidence that their lives were tangible and genuine. They paid for their freedom with the loss of their only safe anchorage in an uncertain world—their unquestioning faith in God.

In 18th century England, the Industrial Revolution followed the Age of Reason. Life was exciting for inventors, entrepreneurs, and businessmen, but it displaced masses of people previously engaged in agriculture who left their rural homes to find jobs in the booming cities. Urban factories designed for specialization made the large-scale production of goods possible. An expanding population demanded more and better manufactured products and this emphasis led to an exploitation of the factory workers—men, women, and children. Many workers were forced to spend long hours in factories earning little pay for performing repetitive tasks on assembly lines. In time, the workers in the factories lost their human identities. Their bleak lives gave "invisibility" an ominous new meaning.

The conditions at the time of the Industrial Revolution created new social classes—owners, managers, and workers with little responsibility towards each other. Exploitation of factory workers was accompanied by the dislocation of families crowded into tenements. Workers were considered little more than human robots. The person-to-person relationships that had held people together under feudalism no longer existed during the Industrial Revolution and the need for human interrelationships remains largely unmet today. In the United States we pay the price for the lack of social connectedness by the proliferation of street gangs that attract many of our teenagers. These gangs offer their members identifying jargon, clothing, territories, strong internal loyalty, and even the togetherness provided by gang warfare. Such activities represent nothing less than a return to a kind feudalism in miniature.

I request the reader's indulgence while I dramatize the lost feudal personal relationships by asking the following question: What would one think today of an employee who vowed lifelong allegiance and sank to his knees before the president of a business corporation who promoted him to a middle management position? Under feudalism, this scenario was customary when a lord awarded a portion of land to a vassal. In the medieval promotion ceremonies, the overlord kissed the kneeling vassal and raised him to his feet. Thereupon, the vassal swore an oath of loyalty and vowed to be faithful forever. I am not recommending that we create similar promotion rituals today. However, our modern profit-oriented society is largely devoid of a sense of obligation and personal loyalty. An impersonal view of our fellow citizens is typical in business transactions and in our civilization at large. In our milieu, we direct our warmer emotions primarily to issues. People, on the other hand, have become abstractions and symbols. This could lead to irresoluble controversies when the human genome is fully mapped and we are ready to apply genetic engineering.

The prevailing climate of anxiety and uncertainty of one's role in the world, precipitated by the need to obtain visibility, has led us to a 20th century movement called existentialism. It is a reaction to the previous view that the single fragment of reality—visibility—could, by itself, give life meaning. Both Christian writers and atheists contributed ideas to the movement. Existentialists denied that people required visibility to exist. They proposed, instead, that existence precedes essence, that is, first one becomes a person, then one becomes a particular kind of a person. Existentialism reversed Bishop Berkeley's "I see a tree therefore the tree exists" to "It is a tree and since I'm human it looks like a tree to me." Its theme consists of the idea that freedom and the lack of given rules create the need for us to be responsible for our actions. A controversial writer and philosopher, Friedrich Wilhelm Nietzsche, who is discussed later, has been viewed by some as a forerunner of existentialism.

Existentialism gained considerable acceptance in Europe in post World War II France. Jean-Paul Sartre, (1905-1980), French playwright and novelist, became a leader of the existential movement. In a popular novel he described humans as lonely beings who were burdened rather than elevated by their freedom. He portrayed the current generation as drifting aimlessly through life. According to Sartre, people's anxiety could be viewed more accurately as anguish caused by the almost impossible task of giving meaning to a meaningless world. Existentialism offered an alternative to the notion that people must validate their lives by exhibiting their visibility. Instead, the new thinking suggested that people could give their lives meaning only by accepting responsibility for their own fate. Some people viewed existentialism as an elitist concept and existentialists failed to get their message across to the man in the street. People were not ready to accept the idea that they were responsible for their own problems but, instead, continued to rely on visibility to demonstrate that they were unique individuals. Today, responsibility still takes a second place to calling attention to oneself. We now experience a revival of "seeing is being."

Some scientists believe that the approach to consciousness is almost solely through vision. In his book, *The Astonishing Hypothesis:The Scientific Search for the Soul*, (1994), Francis Crick, the 1953 Nobel laureate co-discoverer of the structure of DNA, argues that consciousness, "the essence of humanity," as he sees it, must rely on science to explain it. He recommends the use of visual neurobiology and psychophysics as approaches to reach consciousness because our neocortex has adapted vision as a primary resource for gathering information. Consciousness, however, goes beyond information and awareness. An important aspect of the essence of being human lies outside the reach of physiological vision and that applies equally to *soul* whatever meaning one may give it. Inner experiences rather than visual processing lead more directly to the soul while seeing does not always lead to inner-experiences. The matter is complex but if everything we could ever learn from science were absorbed by us, something would still be missing. Perhaps in addition to whatever else occurs this something plays a role in the meaning of life. J. J. Hopfield of The California Institute of Technology in his review of Crick's book, in *Science,* (1994), writes, "The book is a heroic attempt to wrest consciousness from the minds of philosophers and place it in the hands of scientists." He referred to those scientists, I might add, who still believe in Bishop Berkeley's and Descartes' claim that physiology is the sole repository of reality.

Esse est precipi has invaded the halls of academia. Among college professors it is called "publish or perish." Excellent teachers who do not gain visibility through publishing professional books and articles find themselves in danger of losing their jobs. The use of violence such as detonating bombs or hijacking planes is a less academic way to gain visibility. However, it quickly creates visibility by capturing media headlines. Our contemporary violence is, in part, an expression of frustration that the utopias promised in the Age of Reason have failed to materialize. People are tired of waiting for the good life the Enlightenment promised and have fears that they have been misled.

The descendants of the European idealists who brought about the Age of Reason are now returning in ever larger numbers to the very authoritarianism, fundamentalism, and ultra-conservatism against which their 17th century forefathers revolted. They demand to know where they belong without having to demonstrate that they exist. Nikki R. Keddie, professor of history at the University of California at Los Angeles makes the point in a postscript in *Contention*, (1993), that resurgent fundamentalism is a modern trend which stems from widespread frustration with the failures of westernized nations to satisfy basic human needs. She has spiritual needs in mind that may be beyond what the Age of Reason was able to provide. Later in the book we shall try to identify what these needs consist of.

The Age of Reason encouraged democracy in governments and championed people's right to live in dignity, liberty, and freedom. But it failed to integrate visibility with important but less tangible aspects of life. The Industrial Revolution stimulated ingenuity and inventiveness. Nevertheless, it depersonalized and dehumanized many people living in urban Western Europe and the United States. It led to wide spread unemployment during periods of economic stagnation caused by the replacement of manual labor with more efficient time-saving machinery. Can-do humans with nothing to do soon feel useless and become emotionally disturbed or embark on a career of crime. Some of our present dilemmas are rooted in the ground that the Age of Reason and the Industrial Revolution prepared for us. Notwithstanding the benefits we received from the Enlightenment and from urban industrialization, we have lost the feeling of human interconnectedness. The feudal system with its unbreachable class distinctions of nobles, clergy, and serfs still managed to maintain a togetherness of workers and aristocrats, all ruled over by an omniscient God. The interdependence of people created feelings of loyalty and obligation. In the modern world loyalty, obligation, and belongingness fail to exist within our social fabric. Today, the best examples of belongingness can be found in ethnic, tribal, and racial chauvinism, and in the brotherhood of international terrorism. Something new must occur within our society if we are to avoid transmitting this chauvinism into the Age of the Genome.

Existentialism called attention to the rootlessness that accompanied freedom from obligation. The theme had an impact on a number of authors who wrote on mental health. The book, *Existence*, (1959), edited by psychoanalyst, Rollo May, claimed that after people in Western civilization freed themselves from hunger and fatigue thanks to the use of agricultural machinery, they ran headlong into boredom and meaninglessness. The authors suggest that in order to transcend the existential vacuum—the emptiness in their lives—people must accept personal responsibility for their unsatisfactory contemporary condition and build a new liveable society as I have maintained throughout this book. Somehow the chance to build such an improved world has seemed to slip through our fingers. No doubt there will be people in the next century convinced that we should use genetic engineering and redesign ourselves as the only way to escape from our present insoluble dilemmas.

The Age of Reason taught us that we have rational minds. Existentialists maintain that we are responsible for ourselves and our fate. We must try to integrate these ideas against a backdrop of indoctrination on political, health, financial, and religious issues given to us by aggressive people with their own personal biases who utilize the media to create visibility for themselves. Our own need for visibility increases our susceptibility to their persuasiveness. Television personalities become real people in our living rooms in front of our eyes. We may view them as modern knights of the media unencumbered by any traces of feudal obligations. After the pictures on the screens fade, the TV personalities remain within us haunting us to do their bidding. The malignant spirits feared by some indigenous people have reappeared in the guise of some media personalities. They use sensationalism in communicating that startles, shocks, and thrills while it misinforms. In spite of warnings against portrayal of violence and clamor for restraint, the violence which continues on the television screens represents a fragment of reality that becomes all of reality on the crime-ridden streets of our major cities. Mapping the human genome will do no more than add another disturbing fragment to the already indigestible mix of contradictory forces that rule our lives.

As has been suggested when reality becomes fragmented, a loss of connectedness, coherence, and interrelatedness of events results. It isn't surprising, therefore, that people in the modern world tend to reverse the logical order of events. We confuse the territorial crowing of the rooster with the laying of the egg, exhibiting accomplishments with achieving them, advertising with manufacturing, and showing with creating. We want to hear the applause before the performance, enjoy the benefits of work before we complete it, spend our money before we earn it.

We can count on this: media personalities will be out in full force in the Age of the Genome to convince their audiences how gene technology should be applied on terms they will suggest. The mind control of feudalism will return in changed guise in the Age of the Genome disseminated by means of telecommunication. By sight and sound, media personalities will try to bend the minds of their viewers to conform to their wishes. Many viewers will become unwitting mental captives, television serfs, a "herd of sheep" as Nietzsche would have called them, in spite of the self-determination which the Age of Reason fought hard to win for us. The human genome may assume the mystical aura of the Holy Grail which in the Middle Ages inspired Christian zeal, King Arthur legends, Celtic myths, and played a role in fertility cults. Thus mystified, scientists conducting genetic engineering would be able to discourage legitimate criticism of their goals.

Don Quixote, hero of the book by the Spanish Renaissance novelist, Cervantes, drove his spear into windmills in his effort to battle the wrongs of the world. The book highlights the tragedy of idealism throughout human history. Since Cervantes wrote it in 1605, isn't it sensible to ask ourselves, "In spite of exposure to existentialism and to the additional power we have gained from our modern technological breakthroughs, are we coping with an unchanging and unchangeable human nature?" If the answer is "yes," many will claim that we can create a better world for humankind only by redesigning our human genome. The temptation to redesign ourselves genetically may not be easy to resist because the

Age of Reason failed to live up to expectations. It is time that we look further back to before written history for clues that may help us gain some insights into human nature. In the next chapter we shall use the largest perspective available to us, to see how human nature reflects the nature of the universe and how both would be diminished if we redesigned human beings in an attempt to achieve what the Age of Enlightenment failed to accomplish.

CHAPTER FIVE

OUR COMPULSION TO JOUST WITH NATURE

The laws that govern the physical nature of the universe play a role in shaping human behavior. We shall better be able to understand certain aspects of ourselves if we learn more about our interdependence with the nature of the universe. From the Big Picture view our genome represents a product of the human-cosmos partnership. In this chapter, we shall try to fit gene technology into a larger picture which includes not only genetic engineering but our perception of our world as a whole.

Daniel Pendick, (1993), whose articles appear in scientific journals, may have had this in mind when he wrote, "At its grandest level, science is the ongoing attempt of humans to make sense of the universe." Without making sense of the universe, it is difficult to understand our own lives. If we confine our lives to the small picture view, we will inevitably fragment reality. Long before Pendick, the Roman philosopher, Lucretius, (99-55 B.C.) in *De Rerum Natura* came to a similar conclusion. "Let man's first study be the knowledge of the nature of the universe," he advised the people of his time. In order to try to understand the nature of the universe we must take into account that it represents an abstraction that includes ourselves, our culture, and the extent of our scientific knowledge.

The universe exhibits characteristics discovered through scientific efforts and these discoveries, themselves, represent aspects of its potentialities. It has laws, dynamics, and physical properties that impact on all things from the smallest subatomic particle to the largest constellation. From the view we take in this book one of the characteristics we stress as noteworthy is that the universe produces dynamism—a push that makes things happen. From its beginning to the present moment, the dynamism of the universe continues to transform its raw potentialities into happenings that create an ever-expanding reality. Within this context we can certainly agree with Bishop Tutu that manifesting the glory of God gives human life meaning; great human accomplishments may be viewed as manifesting the glory of God. Characteristics of the universe—harmony, struggle, and dynamism—are reflected in human nature (T. C. Kent, 1989). The dynamism of the universe is reflected in vital energy that converts one state into another at the molecular level. This is too narrow a view of human life since it fails to explain synthesis and transcendence, which we shall discuss in later chapters. First we shall explore the significance of nonpotentiality—negative reality—that remains untouched by the dynamism of the universe. Since nonpotentiality cannot cause anything to happen, it may seem strange that, nevertheless, it plays a meaningful role in how we think and act.

A brief review of the history of the universe as seen from our human perspective may help us appreciate the role that nonpotentiality plays in our lives. We begin speculations at the opposite end of a continuum from where the British zoologist, Richard Dawkins, and others championing self-interest started theirs. It leads us to different conclusions from those Dawkins arrived at when he wrote that morality in human behavior should be considered as enlightened self-interest. In the preface of the 1989 edition of his book, *The Selfish Gene*, he stated that he did not wish to focus on humankind or on the individual person but rather "take the gene's eye view of nature." In contrast, I prefer to explore conclusions we may reach from a cosmic view of the gene. I would rather think of the gene as an actualization of the potentiality of the universe than view the universe as an actualization of the potentiality of the gene. The latter might qualify as an example of our penchant to reverse the logical order of things that I referred to in the previous chapter.

From the view of science, the Big Picture starts with the birth of the cosmos. Cosmologists are not agreed, but most of them estimate that the primeval explosion, sometimes referred to as the Big Bang, occurred approximately 15 billion years ago and created our still-expanding universe. This view of the beginning of the universe is known as the evolutionary hypothesis. An alternate hypothesis proposes that instead of an explosion, the universe always existed and that matter is continually created in one place as it is destroyed in another. This view is called the steady-state hypothesis. The majority of cosmologists accept the explosion hypothesis but the steady-state versus the evolutionary hypothesis has not yet been settled to everyone's satisfaction.

Cosmologists think that stellar dust and gases formed the solar system. Scientists estimate our own earth as approximately 5 billion years old and fossils dating back to about 3 1/2 billion years have been found. If we placed the length of time life has existed on earth on the dial of an imaginary clock, hominids—the earliest human-like types—entered the scene about 2 minutes before midnight with a brain roughly one third of the size of modern humans. Now our species has the largest ratio of brain weight to body weight in the animal kingdom with one exception—the Capuchin monkey of South America. Nature always seems to find a way to keep us from becoming too conceited.

In spite of the size of our advanced brain, in terms of life on earth, we are mere infants. Perhaps, because of our youth we might be forgiven for some of the ruckus we have caused on earth with this enlarged brain. It might occur to us that gene technology may well be a toy more appropriate for children older in geological time than present day humans and, perhaps, a few million years more mature.

Because we are in some ways like precocious two-year-olds, we rebel against the idea that certain things are forbidden us. Forbidden edibles, for example, become more attractive whether grown in the Garden of Eden or stored in our home refrigerators. Throughout time, various cultures have had taboos that prohibited certain kinds of behavior and threatened supernatural punishment for those who disobeyed. Some taboos were related to marriage, to eating specific foods, or touching or naming sacred objects. Those who broke taboos were threatened with severe punishment by supernatural powers since the command, "You're not allowed to do it" sometimes provides the incentive to disobey. We can expect this oppositionalism in human nature to surface in the Age of the Genome as it has in the Computer Age when pranksters created havoc by inserting computer "viruses" into government communication systems some years back. Imagine viruses of similar destructive tendencies surreptitiously introduced into the machinery of gene technology! What satisfactions could anyone gain from creating such mischief? One answer could be that it enabled the pranksters to think of themselves as powerful. Imagination like muscles seeks exercise. Furthermore, in their own way, these mischief makers would be activating unpleasant potentialities of the universe—a strong human temptation.

Young children imagine that they accomplish in actuality what they can picture in their minds. Their eidetic imagery enables them to believe that people in stories that are read to them actually exist. Adults may exercise their imaginations by reading and thinking. They can easily think of things that are impossible to do. Certainly, perpetual motion machines cause no problems in science fiction literature. Few things could be handier to own than a perpetual motion machine that would work continuously without fuel. In imagination, complicated thinking isn't required to figure out how it would work. Its wheels would just keep spinning on and on forever. But the laws of nature make that machine an impossibility. The philosophers of ancient times might have predicted that humans would invent perpetual motion machines. It is unlikely they could have imagined a gene technology that would cause mice and pigs to produce medical products for the benefit of humankind. In imagination, sometimes the impossible seems more likely than the improbable. The unbreachable difference between them is that improbable events respond to the laws of statistics whereas impossible events are built into the nature of the universe itself.

Dennis Diderot, (1754), the French material philosopher, who enlisted the leading French talents in the Age of Enlightenment to help him produce an encyclopedia, disposed of happenings with, "Matter acts because it exists and it exists in order to act." I like to put it: happenings occur because the universe is dynamic—pushing things to happen—and they are able to occur because their reality conforms to the laws of permission. Impossible things are excluded forever from actualization by the laws of denial. Never can anything travel faster than the speed of light. Its speed is a universal constant at 186,000 miles per second. Never can the amount of energy in the universe be increased nor decreased. Never can energy which has been used, return to its original state. Even today it isn't easy to grasp the idea that when a stone lies at the edge of a cliff it has potential energy. After someone pushes it off the cliff, its potential energy converts instantly into kinetic energy, the name given to bodies in motion. When the stone rolls to a halt at the bottom of the cliff, it retains its previous energy which then becomes static. Energy and reality are inextricably linked and that is why reality, as I present it in this book, cannot be destroyed. Energy may be seen as the universe's resource. The dynamics of the universe provide the push, actualizations of the universe's potentialities give it a direction for expression, and reality results. Like energy, reality cannot be destroyed, as I have pointed out in chapter 2. This is an important concept to keep in mind when reading the chapters that follow.

The universe's *usable* energy though vast is, nevertheless, limited. This, in turn, limits the number of possible actualizations of the universe's potentialities. At first, this doesn't seem like anything we need to worry about but many environmentalists view diminished, irreplaceable usable energy as a serious problem. In the next century, environmentalists are likely to draw genetic engineering into the arena of their concerns. This might provide us all with additional awareness of our responsibility to conserve energy. We know that whenever energy is converted from one form to another, less usable energy remains in the universe. No engine can ever run at 100% efficiency because the energy put into it dissipates into heat caused by unavoidable friction. "Entropy" is the name given to the measure of the amount of energy no longer capable of conversion into kinetic energy, as when the stone we mentioned took its last bounce at the bottom of the cliff.

Entropy is nature's *enfant terrible*. Its definition includes the amount of loss created by useful potential energy rendered useless when it becomes static energy. An example may be seen when a cup slips out of one's hand and falls to the floor breaking into pieces. The pieces will scatter in a random, disorderly manner. The disorder caused by entropy is irreversible. No matter how many times one picks up the fragments of a cup and drops them again, they will never rearrange themselves back into a cup. Avoiding entropy is one of nature's nonpotentialities. We can blame entropy for the endless job of keeping our house clean, for the wear and tear on our cars, and for feeling tired after we've jogged. Yet most people seldom give entropy a second thought.

Humans, aware of their own lack of omnipotence, want to view their divinities and heroes as all powerful just as young children like to think of their parents as omnipotent. Characters in myths, legends, fairy tales, gods and goddesses traditionally are given the task of ridding the world of nonpotentiality. Pegasus, in Greek mythology, was a flying horse. King Midas of Macedonia—as the legend goes—shrewdly outwitted nonpotentiality by turning everything he touched into gold. Moses drew water from a rock. Buddha walked in the air. Muhammad ascended to heaven on a winged horse. The Hindu mystic Ramakrishna conversed with a stone statue. The fascination provided by escape artists and magicians stems from their seeming ability to perform impossible feats. Even those who know full well that legerdemain is an illusion enjoy watching fellow humans seemingly overcome nature's laws of prohibition.

Napoleon who claimed that nothing was impossible might have been closer to the truth had he added, "in the human imagination." With little effort and no fanfare, imagination easily overrides the laws of denial. As it soars above nature's restrictions, human imagination treats improbability and nonpotentiality with equal indifference. However, imagination can function only within the boundaries of people's awareness of the world around them. Awareness of consequences will become the key to whether or not people will put the future of humanity into the hands of responsible geneticists and law makers. Stimulating imagination became the road to power in the 20th century. Few of us are aware of how precarious our seemingly all powerful imagination, immune to the laws of denial, can be. It can easily be channeled into a self-destructive direction. In the Age of the Genome, imagination may become a victim of popular demand to make all things "real" by encouraging concrete thinking that excludes intuition.

"I think and therefore I exist" Descartes decided, and his words are true only in that whatever is imagined exists as an event within the human mind. Seeing creates reality Bishop Berkeley stated. When this idea is applied to seeing with the mind's eye, it becomes true since imagined things do create subjective reality. The poet, William Wordsworth, realized this when he wrote the line, "Imagination...is but another name for absolute power." We cannot override the laws of denial in actuality, but we can do so easily in our dreams and thoughts. The laws of denial don't count, as in our imagination we resurrect the dead, turn time backwards, banish entropy, and build perpetual motion machines effortlessly.

Human imagination is truly nature's miracle. Light travels at a speed of 186,000 miles per second and takes one hundred million years to traverse our galaxy. I have travelled the distance in a few seconds when my eyes strayed from the screen of my computer to the ceiling in my study. Even imagination in the Age of the Genome may become redundant, unnecessary, and out-of-date if replaced by genes that limit its focus exclusively to practical applications.

The brain of *Homo habilis* overcame the limitation of our anatomy over a million year ago when our human-like ancestors first used stones as tools and sticks to extend the short reach of their arms. Our ability to circumvent human limitations aided by our imagination gave birth to our can-do human spirit. With this spirit, molecular biologists in their laboratories work with their instruments to map the human genome day after day sequencing and linking the positions of genes' specific known characteristics. Almost monthly we read of breakthroughs that enable geneticists to map genes faster than had previously been anticipated. Information on gene technology rapidly becomes out-of-date.

A first step on the agenda of gene technology must be to identify and eliminate inherited diseases. After this we may place an uncertain foot on the next step of the ladder of genetic engineering. That step may be the use of gene technology to prevent those mental disorders identified as having genetic components. I have referred to eugenics previously as practiced in the earlier part of this century. Its purpose was to eliminate mental illness, retardation, and crime by sterilizing people considered defective. The same reasoning may again be used in the Age of the Genome to improve humankind. There is quite a difference, however, between eugenics and gene technology. Sterilization as practiced in eugenics prevented the birth of offsprings who had the potential to inherit their parents' genetic problems while gene technology would try to ensure that children born would be endowed with minds and bodies genetically altered to be free of defects. Just as abortion issues are controversial now, this is certain to provoke debate in times to come once the capability to redesign humans has been mastered.

Still another step in the future application of gene technology may consist of attempting to alter and improve people's physical appearance, or should I have mentioned this step first in view of vanity's persuasive power? An almost irresistible pressure may be placed on molecular biologists to achieve this goal. The often quoted 17th century French epigrammatist, La Rochefoucauld, wrote: "Vanity can make us do more things against our better judgment than reason can."

Let us look ahead and try to envision some of the hurdles scientists may have to overcome before they can influence human behavior by means of genetic engineering—a mindboggling undertaking because of the large number and variety of different genes and enzymes which play roles in shaping behavior. This situation appears more complex when we consider that specific genes, known as messenger genes, activate other genes that help to produce certain human traits. The difficulty of ascertaining genetic versus environmental influences on personality will continue well into the next century.

With some exceptions, genes are generally team players and mutually interdependent. When geneticists attempt to change the human personality, they will have to deal with gene configurations instead of just individual genes. But the difficulties don't stop here. Confounding the problem for many years to come is the fact that identical environments do not affect people in the same way; therefore, preventing illnesses by extracting and replacing genes identified as contributory to depression, schizophrenia, compulsions or aggressions will be a monumental undertaking with uncertain results. The then old fashioned psychotherapy of our 20th century may be much less risky and more practical to use. However those able to perform near miracles are not often deterred by the fact that there is no need for them.

No doubt we shall see a determined government, aroused religious leaders, wary scientists, and a crusading public demanding restraint in genetic experiments with the human psyche or as some would prefer to call it, the human soul. Again the cry "We shall overcome!" may be heard in song and verse as so often before in human history. In the next century the words to the song might be, "we shall overcome the urge to do genetic engineering!" As far back as 1974 Halacy wrote, "The genetic revolution may produce a new species. The question we should ask is whether such a new species would flourish in quantity or in quality." Halacy then raised the possibility that *Homo futurus*, might not survive genetic engineering but pass away as 98 million earlier species of life already have.

One of the promising aspects of gene technology is that it can reduce entropy by replacing mechanical engineering with bioengineering. Molecular biologists are devising ways to extract nonpolluting power from living organisms that will be able to replace our present use of fossil fuels with an immense saving of effort and time. In the Age of the Genome, useful enzymes will be grown by bacterial colonies. Microbes will consume our ever-increasing waste products and in the process, produce methane gas to fuel automobiles and locomotives. Molecular scientists will develop biological sources of usable energy. This use of genetic engineering will bring with it a new industrial revolution fueled by bioenergy. After this biological breakthrough has occurred, a new potentiality of life's forces can be harnessed for the beneficial uses of humankind. But we can safely predict that soon thereafter the human mind will be drawn to the next item on a never-ending list of challenges. The dynamics of the universe flowing through the human imagination will always search for them.

It was in 1953 while they were at Cavendish Laboratory, Cambridge University that James Dewey Watson and Francis Harry Compton Crick found that the DNA molecule was composed of a double helix. Recently, Dr. Crick's published *The Astonishing Hypothesis:The Scientific Search for the*

Soul (1994). Crick writes, "The Astonishing Hypothesis is that you, your joys and your sorrows, your memories and your ambitions, your sense of personal identity and free will, are in fact no more than the behavior of a vast assembly of nerve cells and their associated molecules." However, from my point of view, I see my joys, sorrows, memories, ambitions, my identity and free will as potentialities of the universe and myself merely as their activator. Crick starts at the neuron level and arrives at the Big Picture. I recommend, as mentioned previously, that we start with the Big Picture which leads not only to the neuron but to every other destination in the universe. Crick travels the same route from neuron to universe but does so from opposite directions. This causes us to arrive at different conclusions. Neural networks, brain responses to external stimuli, the reflections and interpretations of such stimuli are not singly or collectively responsible for the "soul," whatever that might mean. To think so is to take the small picture view. Yes, we have a role in building the universe by our thoughts and acts. But our capacity to do so, the element which reductionists tend to ignore, stems from the potentiality of the universe rather than from anything that exists apart from it isolated within the human brain. Although our brains help make us the most gifted actualizers on earth we do not create the primary stuff—the potentialities inherent in the universe that allows what we do to happen.

Friedrich von Schelling, a 19th century German philosopher who admired human creativity and art put it better than Crick when he wrote, "God, the Great Artist (I add, call "Him" or "Her," or if you wish, "the potentiality of the universe")...is still creating the universe *through* us." For example, it is incorrect to think that humans create music within their brains out of nothingness as if by spontaneous combustion. Instead, a composer by means of his creativity actualizes a potentiality of the universe which his composition expresses. Were it not for this potentiality no composer could conceive of a melody, a tone, or an orchestration. As they play the music to the audience the musicians manifest the potentiality of the musical creation the composer actualized. To view this as reflecting the power of God makes more sense than thinking of it as created unaided within an isolated fragment of reality—a fragment called *Home sapiens, sapiens*. Even the earliest humans who used sea or coconut shells to scoop up water from a lake or river knew what we sometime tend to forget, that *for something to come out, something must go in.*

Viewing the human brain as creating internally what it can accomplish ignores the laws of denial although those who do so will nevertheless maintain that these laws are valid. This seems to be a currently popular game played by some researchers in neurobiology and psychoneurology. It is a dangerous game to play in the Age of the Genome since it consists of planting one foot on a fragment of reality while the other foot rests on another fragment that is drifting away into a different direction. Recognizing that there has to be input from the universe for things to happen represents the Big Picture view. Ascribing the origins of phenomena to our internal mental equipment, thus bypassing the mystique inherent in the universe's potentialities, can only result in an egocentric use of gene technology. This is because it tends to make us forget that we are not omnipotent and those who forget this even briefly in their rush to assure themselves that they exist *a la* Descartes may ignore possible undesirable consequences of their enterprise for humankind.

The laws of denial—the *you can't do* laws—will remain eternally unconquered. As long as our species survives, they will remind us that we lack omnipotence. For this we owe gratitude to the laws of denial which contribute to the development of our human ingenuity by challenging us to outwit them. Should these challenges require us to view life as a struggle? Or is it more accurate to see harmony in the grand scheme of nature, with struggle playing only a secondary role? In the next chapter we shall examine the arguments that favor each of these views.

CHAPTER SIX

THE ROLES OF HARMONY AND STRUGGLE

Our emerging philosophy of gene technology will in part shape our perceptions of harmony or struggle as the dominant forces in nature. This philosophy will serve as the foundation upon which we will build our future world. Therefore, it behooves us to give some thought to the roles of harmony and struggle before we begin to use gene technology to redesign significant aspects of our lives.

I will use the word, *harmony* here to mean an equilibrium that leads to an underlying serenity in the universe and *struggle* as a cosmic imbalance that creates challenge and danger in life. *Webster's New World Dictionary,* (1988), illustrates the use of the words, *to struggle*: "to make one's way with difficulty as a *struggle* through a thicket." A view representing harmony, on the other hand, sees the same journey as devoid of struggle. We find that cooperation and symbiosis as well as struggle and strife play a role in the survival of almost all forms of life. The question of which of these contributes most to human nature requires an answer that takes us beyond individual opinions and even culturally-inspired attitudes. Let's examine the roles of harmony and struggle from human perspectives and to do so retrace our steps back to *Homo habilis* or other early progenitors of modern humans.

Over the period of approximately two million years that *Homo habilis* evolved into the modern human being a gradual change took place in how humans perceived the world. Instead of merely observing that changes occurred in their environment, as time went on they began to ask themselves why things changed and how that happened. Early in human prehistory, hunters and gatherers must have wondered why the sun and moon didn't fall to earth and what caused the regular cycles of day and night. They watched the stones they tossed into the air return unaided to earth. They felt the wind on their bodies and saw it cause bushes and tree branches to sway. They observed rain fall from the sky and ocean waves roll rhythmically. They may have felt the regular beat of their hearts and realized that there was rhythm within themselves as well as in nature.

Our ancestors observed that plants grew from out of the earth and that they shared birth and death with animals. As some people in isolated parts of the world still do today, they must have reasoned that a variety of spirits inhabited all things: trees, stones, animals, and people, and these spirits caused things to happen. Our ancestors observed that everywhere predators survived by killing weaker animals for food and must have been aware that struggle existed in nature as well as the harmony they saw all around them. When they witnessed catastrophes such as earthquakes, droughts, floods, and forest fires caused by lightning, they were reminded that nature could be terrifying as well as nurturing.

Thoughts of a spirit world must have occurred to the Neanderthal people who as long as 40,000 years ago buried their dead in shallow graves with their flint tools beside them. Lifelike paintings and drawings of woolly mammoths, bison, wild horses, and reindeer still can be found on many cave walls in Spain and France. Anthropologists speculate that these sites may have served as centers of religious rituals. The artists were Neolithic humans who chipped well-designed stone tools 20,000 to 30,000 or more years ago. They too must have believed spirits inhabited the animals they portrayed. People as well as animals exhibited affection, courage, shrewdness, and fear. All natural phenomena even rocks and wood were thought of as inhabited by spirits, a belief known as *animism.*

About ten thousand years ago, agriculture gradually replaced foraging and supplied people with surplus food that they could store for future use. This provided them with increased leisure to try to understand what caused things in their environment to occur. They believed that supernatural beings, usually gods with human-like qualities were responsible for all events that happened. Not all gods lived in harmony with each other and as with humans, strife and jealousy existed among them. Once humans had time to think, the world seemed much more complicated than it had appeared to our earlier ancestors who may never have thought much about harmony or struggle.

In 2000 B.C. the ancient Greek philosophers theorized about the presence of harmony in the universe. "Cosmos" comes from the Greek word, *kosmos,* meaning order and harmony. Philosophers acknowledged that strife existed in nature but when they looked at the starry skies, which they likened to a beautiful ornament, they found harmony better fitting the description of the universe than struggle. According to the Old Testament, the first humans were expelled from the Garden of Eden, pictured as a paradise where harmony prevailed. Harmony, it was predicted, would return to Earth with the coming of a Messiah. Buddhists view harmony as inherent in an all encompassing concept of an Ultimate Reality. Struggle becomes a counter force that results from human ambitions and the pursuit of pleasures. Hinduism describes harmony as the balance of three tendencies: action, stability, and decline. Taoism finds harmony in the interplay of opposites such as in Yang and Yin reflected in earth/sky, day/night,

male/female. Mohammed's teachings emphasize brotherhood, and the Koran offers directions for living that would
replace strife with harmony. It is clear that the modern world
religions reflect the harmony for which people always seem
to yearn.

Eastern views of harmony do not contradict the can-do perception of humans. Religious teachings that lead to harmony
through God do not deny that struggle may occur—inner
struggle at least. All explanations of nature's way take into
account the dynamism of the universe which Taoism describes in beautiful simplicity as a "creative force in nature
that gives humans the role of co-creators." What might
Taoists reply if asked, "Since we are co-creators, isn't genetic
engineering a legitimate pursuit for humans?" Perhaps they
might reply that the things nature creates are not appropriate
for humans to create. They might consider that designing life
is best left to nature.

In the 18th century some naturalists in France, England, and
the United States saw the evidence of harmony in the way
nature controls overpopulation. They noted that when one
species becomes too numerous to insure its survival, nature
restores balance by natural disasters or by increasing the
number of its predators. This led people to view nature as a
benign and all-wise guiding force. But others observed that
many animals have to defend their territories. Males of some
species engaged in a kind of psychological warfare competing
for mates. The food chain that sustained life is by and large
a chain of cruelty and death. To some, an ongoing struggle
for existence seems the most accurate interpretation of
nature's way. Struggle may be seen as the process by which
species that adapt to their environment survive while those
who are unable to do so perish. Charles Darwin published
his theory of evolution in *On The Origin of Species* in 1859,
often referred to as "the book that shook the world." Later
Darwin wrote *The Descent of Man,* (1871), in which he expanded his earlier theories. Darwin considered the survival
of the fittest through physical struggle as the most important
aspect of natural selection. Survival was a species' reward

for competing successfully with other species for the same limited ecological niche.

In the aftermath of Darwin's theory, Herbert Spencer, the influential English biologist, philosopher, and writer asserted that the elimination of the weak of a species by natural selection through competition would eventually lead to a state of equilibrium and harmony. He equated successful adaptation with ethics and saw failure to adapt as a moral evil. In essence, Spencer believed that nature attempts to rid itself of its mistakes. This includes eliminating people who are unable to compete. Spencer viewed a nation that helped to keep the unfit alive as doomed to extinction. This caused him to champion *laissez faire,* the idea that a state should not interfere with private initiative in any manner. Consequently, Spencer opposed state banking, state supported education, state charity for the poor, and state support and regulation of housing. He believed even taxes and government postal services were undesirable. One can't help wonder what role genetic engineering would have played in Spencer's philosophy of noninterference with natural selection had he been able to anticipate its future discovery. Would he have placed genetic engineering into the same undesirable category as taxes and postal services or would he have had to revise his view of non-intervention if geneticists demonstrated that gene technology could some day design the most fit?

After Darwin's theory of evolution had gained world-wide attention, some sociologists attempted to introduce the biological principles found in evolution into social reforms. They warned their government leaders not to interfere with the use of competition to eliminate the unfit since, they felt, competition was nature's way of improving life. These proponents of unfettered competition were called Social Darwinists. However, other writers on social issues strongly opposed their views. One of these was the influential Scottish preacher and writer, Henry Drummond, who delivered a series of lectures in 1894 at Harvard University entitled the *Ascent of Man.* He used *ascent* to emphasize his opposition to Darwin's perception of human's descent from other forms of life. Drummond didn't deny that there was struggle for survival but viewed it as only a first step in what he called, "the struggle for the lives of others." The impact of Drummond's insight has been generally lost. His perception of struggle too often fails to enter into the consideration of discussions on whether harmony or struggle represent nature's way. With our large variety of options, nothing in nature suggests that struggle in human life must necessarily be *against* something or someone. A struggle *for* something does not diminish its challenge. Struggle for something is often called striving. However, struggle, striving, harmony, balance—are not mere questions of semantics. Both struggle and striving represent efforts for creative changes and thereby new human potentialities.

Edward Bellamy, (1889), an American novelist went one step further than Drummond when he claimed that many who succeeded in the competitive world of the Social Darwinists were really the *unfit* in terms of what counted most—humanitarianism and compassion. The debate continued as one of Bellamy's contemporaries, Kenneth Walker, a prominent British surgeon and scientist, declared Bellamy's views "shallow." In his opinion humankind had risen from stage to stage mainly by means of competition and ongoing struggle. Jean Jacques Rousseau the political theorist of the 18th century period of Enlightenment extolled the harmonious life of people living undisturbed in natural surroundings. The 20th century Nobel laureate, Francis Crick, co-discoverer of the structure of DNA observed that there was strong selective pressure for cooperation within small groups of people while there was concurrent hostility towards neighboring, competing tribes. He points out in *The Astonishing Hypothesis*, (1994), that, "Even in this century in the forests of the Amazon, the major cause of death among competing tribes in remote parts of Ecuador is from spear wounds inflicted by members of rival tribes." The perception of harmony versus struggle as dominant is confusing because they coexist.

Edward Harrison, a professor of physics and astronomy, wrote in *Masks of The Universe*, (1985), "Some people have the impression that the physical universe is a world of extreme violence...explosions and...cataclysms." However, "the closer we examine the design of the...universe the more we marvel at its harmony and beauty." Einstein proposed similar views. It is not difficult to see how the fixed laws of the universe give cosmologists the impression of order and elegance. Yet, healthy can-do people upset balance every time they change the *status quo*. Ernst Mayr, Professor of Zoology at Harvard found that "the concept of the benign balance of nature" could not be sustained. He points out in *The Growth of Biological Thought,* (1982), that the idea of sustained harmony in the world became untenable when fossil records revealed that many species had become extinct and when geologists learned how greatly the world had changed geologically throughout the ages.

We need not consider the issue of harmony versus struggle from the view of geological ages and extinct species. It can also impact a human in the span of one lifetime. In 1993, a brochure introducing seniors to an Elderhostel art course quoted an 84-year-old woman's list of things she would do differently had she the chance to live her life over. The list began with, "I'd like to make more mistakes the next time." The mistakes she referred to had nothing directly to do with struggle or harmony yet they had implications for both. The woman felt that she had played her cards in life too cautiously. She had tried not to upset anyone or anything. In retrospect she thought that risks associated with struggle would have enhanced her life.

Of course, balance exists in the universe but if balance were primary and humans were to adapt themselves to it, we would have to become accustomed to watching sporting events in which every team's score tied every other team's score all season long. Had harmony existed in the days of chivalry no medieval knight would have wanted to unhorse another knight in an attempt to gain his special lady's admiration. Adolf Gruenbaum, writing in *The World of Physics,* (1987), observed, "The universe around us exhibits striking disequilibrium of temperature and other inhomogeneities." We can be glad it does because without the imbalances and "in-homogeneities" many of the universe's exciting potentialities would have remained forever dormant and our species, *Homo sapiens, sapiens* (the double "sapiens" is correct), would have been among them. Furthermore, the woman quoted in the Elderhostel brochure might then have had no reason for regrets and fantasies of what might have been.

Nevertheless, the concept that balance and harmony exists throughout the universe is easier to accept than one of ongoing struggle. We experience a sense of harmony with nature when we are in close contact with it. There is something satisfying about the idea of harmony that seems to come from a source deep within us. We enjoy the beauty of the sky and our life-sustaining earth with its many hues and forms. We sing and dance to the rhythms of the universe as they pulsate through us. Often at such moments we loosen up and feel as one with all else that exists in the cosmos. But how really inappropriate it is at those times to consider the universe peaceful when nearby a snake swallows a rat, a hawk zooms down to seize a rabbit and, a spider devours a fly trapped in its web. Nevertheless, it is more pleasant to think of song birds as singing from *joi de vivre* than as warning other birds, "This is my territory. Stay out!" We can admire a peaceful scene of grazing gazelles but an ethologist would know that the peace was attained only after vigorous horn butting utilized to establish the dominance of the most aggressive male. Other eligible males excluded from the jealously guarded harem eagerly wait for an opportunity to take over the role of the dominant male. The microscope reveals the Big Picture view as well as the telescope does. The Big Picture view of the universe encompasses the small picture view of viruses and other microbes that infiltrate our cells and subvert our DNA to make copies of themselves at our expense. The T-cells of our immune system that protect us against our ever-present miniature enemies belong to the Big Picture view of the dynamism of the universe acting through life. Wherever we look mostly struggle swirls around and within us, but since we prefer harmony we would rather not think of it.

The idea of disequilibrium is unappealing. We find a picture hanging at an angle on a wall disconcerting. Even in someone else's home we feel an urge to straighten it. We want to see things balanced. Balance is a potentiality of the universe we seem compelled to actualize. Yet it can be argued that we need imbalance to give us our can-do spirit just as the Big Bang required disequilibrium—physicists call a space warp—to make it bang. Could our wish to straighten a picture that hangs askew reflect meaning in human life? If so, when the Big Picture hangs straight, perhaps the work of the universe will have been completed. When the final homeostasis occurs many billions years in the future most cosmologists predict it will bring all things to a complete and final halt. It is interesting that other scientists go on from there to theorize that another universe may emerge from the demise of our own, while still others speculate that there may not be a demise since the universe could go on expanding forever.

Most cosmologists theorize that when the universe eventually collapses into itself and shrinks into a measureless gravitational singularity sometimes referred to as a "Black Hole," its irresistible gravitational pull will suck all things into itself—everything, even including light. Then, perhaps, the can-do challenges derived from imbalances will fizzle into a final frozen balance at last. The internationally renowned cosmologist, Stephen W. Hawking states in *A Brief History of Time,* (1988), that after the universe has become a "Black Hole" it must settle down into a state in which it no longer will be pulsating. After Hawking's book was published additional theories about "Black Holes" were proposed that include "gravitational waves" surrounding "Black Holes" discovered in some regions of our own galaxy. If we accept the view of reality proposed in chapter 2, no gravitation pull will ever erase the history of our universe's actualized potentialities that include you and me and Stephen Hawking and the "Black Holes" themselves. It is a needed cheerful thought in what appears as a dismal conjecture of the fate of the universe.

Let's look at the meaning of balance in the field of mental health. The term, *mentally unbalanced* is another way of referring to psychosis or mental illness. It is a mistaken notion since emotionally disturbed people achieve balance by means of their symptoms while the rest of us continue to struggle with imbalance within and outside of ourselves. Harmony has been defined as *agreement* and this fits schizophrenics better than other people since their delusions enable the mentally ill to agree with every thought that comes into their minds no matter how absurd it may seem to us. Psychotic symptoms permit mental patients to deal with conflicts with which their brain, for structural, neurochemical, or intolerable environmental pressures, cannot cope. Psychiatrists have met "generals", "admirals", "billionaires", and occasionally even a "god" in the locked wards of state hospitals. The easiest way to obtain can-do satisfactions is to develop delusions of grandeur. They demonstrate that the attractiveness of can-do remains even among those who have lost their judgment and reason. A significant amount of all the crime committed world-wide stems from an otherwise unavailable opportunity for can-do.

Mentally healthy can-do people upset balance as they actualize the universe's potentialities. In causing happenings they cannot avoid making the mistakes the 84-year-old woman regretted she had not made. Had she had the chance to live her life over in the Age of the Genome she might fear that a benefactor of humankind would use gene technology to give humans a genetically implanted predilection for harmony in which making mistakes would not be possible.

Harmony evokes scenes of quiet pastures, lotus flowers, and self-effacement. On the other hand, struggle brings to mind competition for ecological niches, horn-butting, and the mapping of the human genome. Were we living in a world of harmony would there be longing for another more exciting world? Quiet pastures alone cannot sustain life. Some people must hunt, fish, or till the land in order to eat. The fear of death is, in part, a fear of harmony—a loss of the opportunity to participate in the struggle required to actualize the dormant potentialities of the universe. Nothing approaches death in unrelieved harmony whether we perceive of it as occurring in Heaven or within the earth. Perhaps, that is one of the reasons normal people fear death—even those who visualize an after-life in Paradise.

Let us look at the question from another point of view. A case for the existence of harmony on earth was made in the Gaia hypothesis formulated by James Lovelock, the English inventor and biochemist in 1970 and Lynn Margulis, professor of microbiology at the University of Massachusetts in 1980. Gaia came of age with the publication of *Scientists On Gaia*, (1992) edited by Schneider and Boston. Gaia maintains that symbiosis, the intimate living together of different organisms for their mutual advantage, cause evolution instead of struggle or competition. According to the theories of Gaia, symbiosis led simple organisms to climb the phylogenetic ladder that eventually brought them to more complex and sophisticated forms of life. The critics of Gaia charge that it introduces a mystical quality into evolutionary theory that, in some way, goes back to the idea that microbes, algae, trees, and rocks are inhabited by the spirit of a "mother earth goddess."

Phil Shannon, long active in the environmental movement in Australia writes in the *Skeptical Inquirer*, (1992), that Gaia brought "a needed self-effacement to an environmentally dangerous, technologically powerful, self-centered species." He refers to humans, of course, but where would one find a species that is not self-centered? One could consider environmental concerns, themselves, as self-centered inasmuch as we depend on the environment for our enjoyment of life as well as for our survival. Nevertheless, Shannon may have a point. "Self-effacement" and the attitude derived from our self-image as humans will be of primary importance to us in the Age of the Genome. Self-effacement is a behavioral innovation in life if not actually in the universe as a whole. The concept doesn't make sense within a Darwinian world order. To understand its significance we must remember that it may represent an aspect of reality and morality.

The question of morality is more complex and controversial since its characteristics fail to follow the usual rules with which we are acquainted. Morality actualizes an unexpected potentiality of the universe so revolutionary that some people view it as the ultimate can-do. First, however, a search for priorities will be useful and this we shall attempt to do in the next chapter. In that chapter we shall use the largest perspectives available to us to see how the idea of a "morality" fits into our perception of the rest of the world.

CHAPTER SEVEN

A SEARCH FOR PRIORITIES

Lois Wingerson combined a career in medical research and writing. In her book, *Mapping Our Genes*, (1990), she states:

> Like Columbus leaving the coast of Spain, the explorers of the human genome cannot see beyond the horizon. They can only imagine what lies there and dream about it. The journey is a little frightening but also very exhilarating. They have persuaded us to pay for it. There is no going back now.

As we stand at the threshold of a new Age of the Genome, we agree that "there is no going back now." Therefore, we must prepare ourselves to go forward and try to anticipate the consequences. Beyond the horizon of the researchers lie the applications of their discoveries and the doubt-provoking questions that the freedom of thought produces. The "dream" to which Wingerson referred could be one of a humankind united in the task of developing an international community of mutually accepting individuals cooperating to employ gene technology for the benefit of all. It could also be a nightmare where contention over who owns the human genome becomes the dominant issue of the next century. If we wish this dream to have a happy ending we should begin now to search for a starting point for our hopes and plans to improve the human condition through the use of gene technology.

In the Middle Ages most people in the Western world were certain that our earth occupied the center of the universe and that a stern, punitive, but if properly approached, forgiving God had created it and ruled over it. As I pointed out previously, people living in those times had no doubts about who they were, why they were on earth, and where they might end up after their death. Their dogmatic religious beliefs and prescribed life-styles were adaptations to the uncertainties and difficulties that touch every human life. Doubt-provoking questions were expediently handled by religious explanations and by blaming Satan. Thus, for the majority of people living in the Middle Ages, all things were accounted for. If these conditions prevailed today, questions of priorities in gene technology could swiftly be answered by the Church and the answers would be accepted as final without question.

We have seen that in the Age of Reason metaphysical explanations and superstitions were replaced by rationalism, science, and freedom of thought. While much was gained for humankind by the "liberation of the spirit," the previous adaptations to life's tribulations no longer served their purposes. Nor has rationalism, as we currently practice it, solved the problems of our times. Molecular biology has raised additional doubts by the discovery of the genetic make-up of life. We must try to resolve these doubts as best we can as we begin to extend the application of gene technology into the next century. Ranking high on our list of priorities for the Age of the Genome we are about to enter is that an effort must be made to regain a sense of community in which people feel mutual responsibility towards each other without resorting to the paternalism, dogmatism, fears, and the superstitions of the Middle Ages.

Because the human genome is the community property of our species, it is more than a little disconcerting to have its application subject to the judgment of those with whom we may not agree. This gives us ample reasons for concern as Wingerson suggests. She points out that we pay for our knowledge of genetic engineering not only with our money but also by permitting research in genetic engineering to take place. The time has come to take responsibility and plan ahead. These plans should not be made by scientists and bureaucrats alone, but should involve many others from as wide a spectrum of the population as possible.

We harbor no illusions that the Big Picture view of gene technology will gain acceptance easily. From Gaia-centered perspectives of Mother Earth, we are of no more importance than the microbes that help maintain the earth's equilibrium. Some have stated that if viruses were to destroy us, the earth itself would not suffer from our extinction. We should note that there are some environmentalists who cast humans into the role of viruses that infect the earth and sicken it. They believe that Mother Earth would divest herself of a burden if our species became extinct. People with these views obviously will have different priorities than those I propose. I believe that our extinction would deprive both the universe and our earth of our contributions that help create their reality, write their histories, and actualize their potentialities. Environmentalists have made it clear that in planning humankind's future we cannot take our earth for granted. I think that neither should we take for granted the contributions made by human existence.

Most creatures other than humans do not despoil the environment as we do but they cannot serve as role models for us because they do not do this on a conscious level. Geneticists must heed environmentalists since the extraordinary physical and biological characteristics of our earth give life on its surface a fragility which must play a role in considering priorities for the Age of the Genome. An earth, as a mother of only viruses, bacteria, worms, and insects, would be deprived of much of its significance. Only humans can perceive the earth's role within the larger universe and, thereby, they can relate to it with greater understanding and appreciation than other living creatures.

The religious philosophy of Taoism correctly describes us as co-creators of the universe. Niels Bohr, the Danish physicist of the first half of the 20th century who was awarded the Nobel Prize for his work on atomic structure saw us as playing an additional role. He believed that a description of a phenomenon must include the measuring system as part of the phenomenon. This means that in selecting priorities for actualizing potentialities of the universe we also must become the measuring system that evaluates them. Some would describe our measuring capacity as resulting from our neuroanatomy and psychoneurology. I prefer to see it as stemming from our religions, philosophies, sciences, superstitions, and even our delusions of grandeur. Some people may regret the last two items on this list but all of them have redeeming features in that they prod us to overcome and circumvent obstacles that could impede the actualizing of our potentialities. Our limitations stem from the fact that we must rely on a Big Picture view as perceived *from our point of observation.* Recognition of the limits of our perception can help us decide where to begin our search for priorities.

I propose that high on any list we should place our desire that human beings will continue to survive. My reason for this differs from those given by sociobiologists who tend to view all living things on the genetic level. The Big Picture view suggests that we have hardly begun to mine the treasury of unactualized potentialities of the universe. For this reason we remain unfulfilled as a species that is gifted with the potential to do more. Let's also admit that aside from these considerations, it's pleasant to feel that in spite of our shortcomings we will be around for quite a while longer especially if eventually we will be able to enjoy a happy human togetherness. Therefore, it should come as no surprise that I propose human survival as our first priority.

The benefits of genetically engineered medical products are more obvious to the public and therefore gene technology for medical use is generally readily accepted. Biotechnology can produce insulin and the cancer-fighting drug interferon alpha. There is even acceptance of the ongoing research producing human hemoglobin and other medical products utilizing pig's blood if it will enable victims of diseases to live longer, healthier lives, and help humankind to survive.

In order to survive we must maintain our can-do spirit. Our ability to use gene technology could not have come at a more propitious time. I have previously called attention to our competitors, the tiny microscopic genetic engineers that could bring about our extinction. Common bacteria that cause human illnesses as certain ear infections and pneumonias are now becoming increasingly resistant to treatment, and may in the foreseeable future, become untreatable by presently known medications. Certain strains of bacteria that cause wound and blood infections no longer respond to the antibiotics available today. Some bacteria on structures called plasmids carry genes that enable them to resist antibiotics. They can share these plasmids with bacteria of different strains enabling the resistance to antibiotics to spread rapidly. The Age of the Genome may gain the distinction of ushering in a post-antibiotic era.

Those who consider our contributions to the history of the universe as important and give our survival the highest priority are concerned about the loss of the effectiveness of current antibiotics. Referring to the new immunity of bacteria to our antibiotic arsenal, Dr. Mitchell Cohen at the Center for Disease Control and Prevention was quoted in the *San Diego Union-Tribune* on February 20, 1994 as stating, "Some common bacteria evolve into wholly untreatable strains. It's potentially an extremely serious problem." He estimated that new drugs necessary to treat diseases resistant to current antibiotics are at least five or more years away. In terms of our priorities we must attempt to design our genetic engineering to be superior to that of the tiniest microorganisms. Therefore, as we wage our war on inherited diseases, we must expand our efforts to include the use of gene technology as a weapon against some of our smallest and deadliest foes. I propose an expanded war against all human diseases as a high priority. Would this also include mental illness? There is impressive evidence to support the existence of genetic factors in certain types of mental illnesses.

I do not think we know enough at this time about how genetic factors contribute to mental illnesses to make decisions on this subject now. It would be wise to view genetic engineering as a last resort in those cases where it might deplete the can-do spirit that helps make us uniquely human. Notwithstanding, if at a later date we gain the skill to alter the genetic factors that lead unequivocally to mental illnesses, we would rid humankind of the miseries and anguish that accompany these conditions. In the future we may have to weigh the question of lifetime imprisonment and the death penalty against redesigning antisocial individuals who, chronically, are unable to resist committing crimes. There may be considerable danger of overplaying our hand in such an effort. But we must balance this danger with a responsibility to alleviate the suffering of victims of remedial genetic abnormalities of all types. In the last chapter we shall consider this question in greater detail. Another consideration in setting priorities takes originality and creativeness into account. Since the dynamic push of the universe leads towards active happenings, it is reasonable to give them a higher priority than passive ones. I do not view relaxing, meditating, or living a quite, thoughtful life as inactivity. Instead I think of inactivity in the sense of not entering into the flow of the dynamics that gives us our unique spirit. I would rank thinking over not thinking, working over not working, creating over not creating. I would place originality that brings novelty to the world high on our list of priorities. Each new happening expands reality within the universe just as the sum of our own external and internal experiences contribute to the reality of our own lives.

There is a special charisma in the phrase, *for the first time in history*, whether it pertains to the beginning of our own lives as celebrated by birthdays or to other new events that take place in our lifetime. For example, an exact copy of a masterpiece of art has far less value than the original painting even if no one can distinguish it from the original. Anything that earns the label *original* creates a special kind of excitement. An event recurring later in the identical or even in an altered form only activates the potentiality of repetition. We see this in records set in sporting events, the discovery of Antarctica, a human footstep on the moon, a new species discovered, or some cosmic event witnessed for the first time. Today's aircraft fly faster than the speed of sound and rise high above the earth's atmosphere. Yet the monument dedicated to flight maintained by the National Park Service was not erected to honor today's pilots. It honors Orville and Wilbur Wright, who made the first engine-powered flight although it only lasted 12 seconds in a plane that was able to rise only inches above the ground. It was a *first* and for that reason its significance outweighs all of the flights that followed. The thrill that we get at the idea of being the first to accomplish a task reflects our role as co-creators in a universe that expands with every "first" that occurs. Of course, not every time a new event takes place is it favorable to us. Nevertheless, the idea of "for the first time in history" inspires awe and, if it is beneficial to us, causes jubilation.

Taking the subject of "for the first time" to geology, we are reminded that climatologists have found that the Earth has been subject to violent climate swings ranging from ice ages to extreme heat. In the last 250,000 years of climate history, the most recent 100,000 seems to have been the only period in which climate remained relatively stable. We were able to build our civilizations during what may have been the only time when climate was sufficiently favorable to allow the beginning of agriculture, the development of culture, and the evolvement of an industrial society. It is predicted by some concerned environmentalists that the benign stretch of stable climate upon which we depend to maintain our quality of life may soon end. Between 2200 and 1900 B.C. thriving civilizations along the fertile valleys of the Tigris and Euphrates rivers, in what now is Syria and Iraq, suffered an abrupt climate change which brought about 300 years of drought and severe dust storms that have been linked to the collapse of that "cradle of civilization."

Nature alone caused this devastation. Environmentalists warn us that today we face a similar threat to our civilizations, but this time the destruction would be our own making. Our Age has contributed a new page to the long history of challenges that we have confronted. These challenges involve undoing the damage we have inflicted on our environment through the careless use of our natural resources, and the pollution of our air and water that are the results of our zeal to be can-doers regardless of consequences. To some extent this trend can be reversed by the use of biotechnology, and I consider further application of bioengineering a high priority.

In the Age of the Genome, industrial chemicals may be grown by bacterial colonies. Some day it will be possible to replace mechanical engineering with bio-engineering. Molecular scientists using genetic engineering may, in the foreseeable future, develop biological sources of reusable energy that could replace fossil fuels. They envision a time when non-polluting power will be extracted from living microorganisms or their sub-molecular parts that would reduce entropy and, at the same time, produce non-polluting energy with considerably less effort and expense than is possible today.

At some future date microbes may consume our ever-increasing waste while producing methane gas that could fuel automobiles and locomotives. This use of genetic engineering could bring with it a new industrial revolution fueled by bioenergy. In essence, a technology that uses genetically engineered microorganisms to perform labor represents a breakthrough similar to the one that occurred when domesticated animals were first utilized for transportation and agriculture approximately ten thousand years ago. This breakthrough changed human life then and increasingly threatened the ecology by over grazing and depletion of the fertility of the soil. The wide spread use of bioengineering may again change human life this time saving our planet from further destruction. Moffat, (1994), reports that within the past few years the mining industry has turned with increasing frequency to biomining that avoids crushing the earth, creating extreme heat, as well as toxic chemicals. This method uses bacteria that separate metals from ores. As more high-grade ore is depleted, biotechnology permits the economical extraction of minerals from low-grade ores.

Agriculturalists have manipulated nature for years utilizing cross-breeding to develop improved varieties of crops as corn and potatoes. With little opposition they have crossed tangerines with grapefruit to produce tangelos. Genetic technology goes beyond this to give wheat, rice, and other grains the ability to resist drought and diseases. Tomatoes, for example, have been genetically altered to ripen longer on the vine while remaining firm for picking and shipping, making them more flavorable and giving them a longer shelf-life at the supermarket. Fruit and vegetables can be given year-around availability. Such breakthroughs could be hailed as a solution to the world's impending food shortages.

In setting priorities, we must take into account what people find acceptable. For example, more than half the people who participated in a survey on public attitudes conducted by the U.S. Department of Agriculture at North Carolina State University found it repulsive to use animal genes to help increase the value of food products for human consumption. A bioengineered gene recently designed to help maintain a desirable texture in frozen strawberries serves as an example.

Molecular biologists are able to isolate a gene in the DNA of an arctic flounder which prevents the fish from freezing in icy waters. Geneticists using today's technology extracted this gene from the flounder and reproduced it in a bioengineering laboratory. They then inserted the gene into the DNA of a strawberry and this delicious contribution to our gustatory joys was altered to retain its texture and taste when frozen and stored. Half the respondents to questionnaires used in the survey admitted they feared toxic and allergic reactions to such bioengineered strawberries. Are these fears justified by the possibility of unanticipated dangers in bioengineered foods? Or will such concerns, in time, join the long list of baseless objections to innovations?

The agricultural biotech industry is expected to develop into a $50 billion industry by the year 2000. Some scientists do not rule out the possibility that bioengineered foods might inadvertently activate a toxic gene in the DNA of a newly created genetically-altered food. Carefully testing the safety of biologically engineered foods before they become available to consumers ranks high on our list of priorities. Still another valid concern relates to the danger of upsetting the delicate balance of nature when we introduce new species into an environment that is not their natural habitat. The accidental escape of the destructive gypsy moth from a scientist's lab damaged our ecology, and the inadvertent import of zebra mussels to the USA aboard a foreign ship continues to cause serious environmental damage to the waters of our Great Lakes. One may ask do we really know what problems we may create when we genetically alter plants and animals? Careful investigation of possible negative effects of all biogenetically altered forms of life ranks high on our list of priorities.

When a food is confirmed as safe and nutritious, objections may subside but distrust at the idea of altering nature with biotechnology may still linger. Over half of the respondents to the previously mentioned survey at the University of North Carolina were opposed to the use of gene technology when it involved the interchange of animal genes, not only because the idea was disturbing to them, but more particularly because they felt it was *morally* wrong. A consideration and understanding of what is meant by "morality" therefore has a high priority for the Age of the Genome.

As seen by the results of the opinion poll, the perception of morality in the Age of the Genome will have a major impact on how gene technology will be put to use in the next century. The conundrum of morality is the subject of the next chapter.

CHAPTER EIGHT

THE CONUNDRUM OF MORALITY

Why does the concept of morality have the characteristics of a conundrum? What qualities of morality make it difficult to reach consensus on its meaning? To consider these questions we must realize that since the beginning of recorded history people have lived in separate groups, families, clans, and tribes. Later, different religions and nationalities evolved. At the same time the world's people shared membership within a single species capable of perceiving abstractions like virtue, goodness, and morality. It is possible that these perceptions could have brought about a desire for us all to live together peacefully and amicably. Sociobiologists may interpret this desire as an instinct to preserve our common genome. I believe this represents an oversimplification.

From reading the previous chapters of this book the reader knows that I think our dreams for peace on earth and our wishes for goodwill to our fellow humans stem from cognition more than from instincts. Whatever its origin, we know that this dream has never been realized. Throughout our history our membership in one species has led to enmity and a denial of the humanness of others as often as it has created feelings of affiliation. Even the ancient Greeks who developed one of the most enlightened civilizations referred to those who were not Greek as "barbarians."

The world's major religions acknowledge our oneness as a human family. Isaiah, prophet of the Old Testament, proclaimed in the sixth century B.C. that the God of Israel was also the God of all people. The Bible acknowledged the unity of all humans through their common creator. Christians believe that Jesus died for all of humankind. According to Islam, God created the universe out of mercy for all people. The Koran declared that making the earth livable for all the world's people is an ideal endeavor. Buddha attacked one of the great barriers to human unity—the idea of inherited elitism. Since class differences separate people, Buddha taught that a person's birth into a family of high rank does not increase the person's spiritual worth.

Some biologists view intraspecies competition as a fact of life. These biologists place much emphasis on the "selfish" gene that fights for the survival of an individual's genetic make-up. Nevertheless, the recognition of our common genetic heritage provides biological support for the concept of the oneness of humankind. This leads inevitably to the conclusion that all members of our species have a joint ownership of our family property—our genome. In order to benefit from this view of our unity we must find a way to minimize hostilities and jealousies that have split our species into fragments. In the past only military conquest could hold large population groups together. Military conquest as well as missionaries often spread their unrelated cultures and religions to distant areas.

The lack of human unity has been explained by ethologists as due to our animal instincts as, for example, competition for dominance, establishing territories, and defending them from encroachment from other members of our species. Among humans, this may have led different cultures to establish "moral territories" that considered those who had other norms and rules as "wrong" often with the implication that they were therefore inferior. Concepts of right and wrong create our perception of morality. Therefore we must replace our vague awareness of our human relatedness with a strong focus on our oneness in humankind. Only then shall we be able to reach a level of consensus about the nature of morality which we must rely upon as our guideline when considering how and when to apply our growing knowledge of gene technology. When the human genome project is completed, a new sense of human unity may lessen the chances of bitter conflicts.

Consensus on what was appropriate existed to a large extent among hunting and foraging societies who had little need for established territories nor competition for dominance. When philosophers entered the scene, after the founding of the city-states, this changed. People began to discuss right and wrong behavior and to argue about what it meant. S. Kent writes in *Farmers as Hunters—the Implication of Sedentism,* (1989), "Sedentism creates new potentials for interpersonal conflicts that cannot be resolved as readily as when groups are mobile." She explains this by observing that after a population becomes sedentary, "fissioning and moving to different areas [to escape conflicts] are no longer viable options. Instead, people with restricted mobility must address intragroup conflicts in other ways than through movement." With our hugely expanded population, how can we regain the communal sense of acceptable behavior that we have lost through sedenism? It will be difficult since hierarchies and other divisions of people into classes, tribes, and nations serve to stabilize societies and often help to avoid internal strife. However, in the world today such division also cause conflicts and rebellions.

Before the world's major religions evolved, traditions were maintained by oral legends. These didn't try to bring about world-wide consensus because they were designed to explain the unknown, to teach, and to establish traditions applicable to the culture of a specific group of people. This would have made a concept of a universal morality difficult to grasp. Therefore, many people willingly accepted the definitions of morality provided by their religious leaders and philosophers. However, Plato in *The Republic, Book II*, maintained that "unless a person, himself, is able to abstract and define rationally the idea of morality...he apprehends only a shadow if anything at all." Plato thought that philosophers were best equipped to see beyond the shadows but that individuals must also work out the nature of morality themselves.

Modern religions differ in some respects in regard to what should be considered right or wrong. Still there is a common thread that runs through the definitions of desirable behavior as, for example, the Golden Rule. Christianity, Brahmanism, Buddhism, Judaism, Confucianism, Taoism, Zoroastrianism, and Islam agree using different words: "whatsoever you would want that people do for you (that is helpful and good) do this even unto others." Even this did not bring agreement because philosophers and people in general differ among themselves about the specific applications of the Golden Rule.

Throughout the years people continue to disagree about what morality consists of. Some accepted hedonism, others adopted situational ethics, or pragmatism as valid bases of morality. Sociologists and some anthropologists continue to view morality as a product of a people's culture. Ethnological studies reveal that perceptions of right and wrong in different societies often fail to agree. This convinces some observers that ethics and morality can only have situational applications. However, from the Big Picture view, this idea represents a hazardous fragmentation of reality. In a world of different cultures drawn ever closer by improved transportation and electronic technology, conflicts on ideas of "right" and "wrong" continue. We can only conclude that as people adapt to conditions under which they live, their cultures will evolve the ethics that will help them to survive.

Jean Jacques Rousseau, the Swiss-French philosopher and political theorist, came to a different conclusion. He believed that humans were moral at birth and would remain so in their natural state. He felt that they are corrupted by exposure to the inventions of civilization such as agriculture, science, property ownership, and commerce. Rousseau's idea of the "noble savage" had considerable influence on later philosophers but was abandoned when explorers found that indigenous people living close to nature were basically not very different from the rest of humankind living under modern conditions. On closer examination indigenous people proved capable of selfishness, jealousy, envy, abuse, violence, and homicide. If they appeared better than people who lived in industrialized nations, it was not because they were intrinsically better, but rather because their environment didn't provide the opportunity for them to engage in advanced civilizations' kinds of "badness."

Some sociobiologists continue to propose that the idea of human altruism stems from the same instincts that cause some animals and insects to sacrifice themselves for the survival of their species. A proponent of the idea of an evolutionary biology, David P. Barash, professor of zoology, University of Washington, wrote with enthusiasm on this subject. In *Sociology and Behavior*, (1977), he declared, "Sociobiology is a whole new way of looking at behavior. It is the application of evolutionary biology to social behavior—that may hold the promise for a greater understanding of human behavior." If this were true, it would greatly simplify the concept of morality. Unfortunately, the opposite has proved to be the case. Barash's animal origins of morality only muddles concepts of human behavior even further. It is true that in some instances one can find a kind of animal "morality" in nature. Rattlesnakes never bite each other while fighting. Antelopes refrain from using their potentially lethal horns against each other. Deer lock horns in battles to establish dominance but don't use them to seriously injure or kill one another as they could. Biologists have found termites that explode their guts to release their contents over their enemies who threaten their nests. Rodents who contract contagious diseases have been known to commit suicide by starving themselves, thereby preventing infecting the others in their burrow. The inference that human morality is a derivative of this type of animal behavior is false. Human imagination can override any of our instincts that would prohibit us from killing another human being.   We only need to read our history books or the daily newspapers to know that in our imagination we can dehumanize other humans and ignore any inhibition we might have against killing members of our own species as, for example, in wars that both sides often wage in the name of "morality."

Some sociobiologists believe that humans can attain peace on earth only after they meld into one super-organism analogous to beehives and ant colonies. Edward O. Wilson wrote in *Sociobiology*, (1980), "a single strand does, indeed, run from the conduct of termite colonies and turkey brotherhoods to the social behavior of man." Such biological-centered views differ substantially from a scenario of humans choosing voluntarily to live in a world community. The concept of community entails cognition and choice while super-organisms imply automatism.

Dogs who risk their own safety to protect their masters from harm act from instinct which is reinforced by training. Such so-called "moral" acts have roots that differ from those properly called "morality."   Primatologists have discovered astounding genetic similarities between humans and apes—blood types, intelligence, ability to recognize signs and symbols, and even a sense of humor. However, the morality proposed by the Golden Rule involves cognition and for that reason differs fundamentally from the involuntary behavior triggered by instincts or social conditioning. Biologically, physiologically, and to some extent psychologically, apes have great similarities to humans. But the moment an ape makes a commitment to a freely chosen morality out of principle, that ape no longer is an ape. Michael C. Corballis, a New Zealand physiologist and psychologist in his book, *The Lopsided Ape—Evolution of the Generative Mind*, (1993), states, "If human uniqueness is not found in consciousness or awareness of self I have argued, nevertheless, that there is a fundamental discontinuity between ourselves and other species."

Sara Hardy, is quoted in *Vignettes; Pitfalls of Evolution (1994):*

> The whole message of sociobiology is oriented towards the success of the individual. It's Machiavellian, and unless a student has a moral framework already in place, we could be creating social monsters by teaching this. It fits in very nicely with...'me first' ethos.

The concept of morality described by sociobiologists is incomplete. It fails to reflect commitments to moral behavior. Morality viewed solely as derived from our past animal behavior deprives it of its meaning.

I have already mentioned that another erroneous interpretation of morality results from perceiving it as solely the product of an adaptive social evolution through culture. This view represents no more than the nurture side of the nature/nurture controversy. The idea that human morality resides in the structure and chemistry of the brain is still another fallacy. This view shares a basic misconception of the locus of human endowment with phrenology. The discredited theory of phrenology was proposed by Lombroso, the 19th Century Italian penologist, who identified human traits by interpreting bumps and bulges on the skull. Lombroso's false assumptions do not differ in principle from the claims of neuroscientists who view the origins of behavioral traits as residing in the brain. It is true that brain lesions, toxins, and electro-chemical stimulation of certain parts of the brain can produce fears, anxieties, and hallucinations. Nor is there doubt that chemical changes, structural differences, amounts of serotonin, dopamine, and other enzymes in the brain significantly affect human behavior. We know that abnormal brain conditions can turn good-natured persons into violent ones who lose their judgment and their capacity to distinguish right from wrong. But in no way does this prove that morality originates in the brain. A radio whose quality of tone is distorted by defective internal components does not prove thereby that it creates the sound it emits within itself. Instead, brain defects and defective radio parts share an inability to respond to external

transmitting stations. Some neurobiologists and neu-
rophilosophers reverse the logical order of things when they
ignore the aspect of morality that represents a human actualization of a remarkable potentiality of the universe. Morality
viewed as originating and residing within human physiology
represents a fragmented view of reality.

Some scientists noting recent discoveries of brain/behavior
correlates suggest that philosophers may have to rework their
notion of social responsibility as they learn the role played by
brain chemistry in making moral decisions. On the contrary,
I think that neuroscientists aware of the Big-Picture view of
interconnectedness of the world's phenomena may wish to
rework their notion of the brain.

Newton's reductionism, the notion that cause and effect
explains all of the universe's phenomena gave way to the new
physics of quantum mechanics. It became further out-of-date
with Heisenberg's principle of indeterminacy and the emer-
gence of a science of chaos, mentioned earlier in the book.
However, today in the behavioral and social sciences reductionism still plays a role in accounting for abstractions like
morality. Scientists are uncomfortable with the idea that
genuine human morality *is not in accord with how we previously viewed nature.* Morality represents a human actualization of a potentiality that, because it lacks precedence,
makes defining it difficult and open to conflicting interpretations.

Two centuries ago the German philosopher, Immanuel Kant,
defined morality as "conduct that requires that nothing be
gained by it." Long before him, Diogenes, the fifth century B.C.
Greek philosopher, according to legends, walked through the
streets of Athens with a lighted lamp looking for an honest
man. Behavior that is not motivated by a desire for gain must
be an evolutionary aberration in a world where life strives
unrelentingly for survival and for the proliferation of species.
A commitment to a morality that, in Kant's view, doesn't
offer rewards seems almost as amazing as the discovery of
genetic engineering itself. Kant's kind of morality cannot

serve adaptation if it consists of behavior from which "nothing is gained." It seems incredible that the idea of morality retains meaning only when it provides no advantage to those who commit themselves to it. Yet, its inability to serve as a means to an end remains an inflexible criterion of its genuineness. People cannot convert morality into something other than that which it stands for nor reshape it to make it acceptable to themselves. Morality does not bend in the wind. It can only break into meaningless pieces. This leaves us no choice but to accept the proverbial saying found in folk wisdom, "good deeds are their own reward."

The human penchant to anthropomorphize inclines us to give human-like qualities to our external world. A century ago the noted German philosopher, Hegel, made an ultimate projection when he declared, "Man is the universe in miniature." Against all of this projection, morality, as Kant described it, stands out as a towering exception. Nothing of ourselves can ever rub off onto this kind of a morality. Some of the content of such a morality can only rub off on us. There is no reciprocity. People who reflect on this aspect of morality may find their confusion about its unusual characteristics increasing even more.

In the middle of this century, J. Bronowski, a distinguished scientist and philosopher formerly at the Salk Institute for Biological Studies, in *Science And Human Values,* (1965), attempted to clarify the paradoxical nature of morality. He pointed out that morality does two different things at one time. It brings people together into societies and also requires that each person within a society remain an independent individual. Bronowski came to the conclusion that no ethic that fails to acknowledge both of these requirements could evolve workable and desirable values nor even permit them to exist. Without them, ethical convictions are likely to deteriorate into prejudices and doing what is right into self-righteousness.

Those who will control experimentation with genetic engineering in the future should take into account that morality which demands freedom also limits it. Moral persons are bound by their own freely chosen rules of ethical behavior. Once they make a commitment to a set of values they must adhere to them unless they change their perception of morality. But this must also be a moral decision and not one for ulterior gains. In genetic engineering we should take into account that morality which demands freedom for the individual also limits it. If, in the next century, gene exchanges between various forms of life can easily take place, the recognition of morality's limitations becomes even more important. More than ever, in the approaching Age of the Genome, we shall need a morality to guide us that meets Kant's and Bronowski's criteria.

Some recent prominent philosophers ignored the Kantian view. Both William James, a pragmatist living in the previous century and John Dewey, the American educator who lived until the middle of this century, promoted the idea of morality as a means to an end for achieving beneficial social consequences. We have already noted that it makes sense to view morality as something that exists for the benefit of society. However, this use leaves the door to morality open equally to humanitarians and bigots. Each could use "morality" differently to support their causes. Those who have demystified morality and reduced it to the level of a tool fail to realize that in doing so they deprive it of its unique asset—independence from serving as a means to an end. The writers of the Judeo-Christian Bible confronted that problem in the Old Testament Book of Job. The illness and grief that befell Job taught him the lesson that good fortune isn't necessarily the reward of ethical behavior. Job did nothing to deserve the suffering he was forced to endure. The purpose of his lesson was to demonstrate that virtue must be pursued for its own sake. When the lesson was learned, the Lord blessed Job with 14,000 sheep, 6,000 camels, 7 sons, and 3 beautiful daughters; and Job enjoyed these blessings until his death at the age of 140. The story had a happy ending.

The writers of the Book of Job shrewdly judged human nature. People then, when it was written, as today, could not believe that good behavior would not be rewarded. After Job accepted morality for its own sake, his fortunes had to improve or few would wish to learn his lesson. Nevertheless, the story of Job illustrates the genuine meaning of morality as a voluntary commitment adopted for its own sake without expectations of rewards. Only human beings are capable of responding to abstractions that seem to disregard the laws of stimulus-response. This capability enables us to experience love, loyalty, and respect for law without ulterior motives. The trial of Socrates is a famous example of a moral commitment to the law. At his trial in Athens in 399 B.C., Socrates was accused of introducing new divinities and teaching rebellious ideas to the Athenian youth. He was found guilty and condemned to death. His friends offered to help him escape from prison but he refused their aid. He accepted the court's sentence and drank the hemlock that was used for capital punishment in his time. His commitment to morality required him to obey the existing law because of *principle* regardless of the consequences to him personally. His was a cognitive decision that overrode his instincts for self-preservation.

In the Age of the Genome, more than in any other time in human history, people must give up the freedom of indiscriminate action to assume the responsibility required to make this a moral world. Herein lies the key that could protect the human genome from exploitation. As stated previously, because morality restricts freedom and cannot serve as a means to an end, it prevents individuals from becoming victims of unfair group-centered social designs. If those in power, in an age where gene technology can be applied to humans, do not adopt Kant's definition of morality, the possibility exists that they might mandate the genetic engineering of others solely to gain control of them. Those controlled would lose their freewill. Obviously, individual freedom cannot exist among people genetically designed to have no alternatives.

Today, many intelligent people believe that moral behavior as a thing-in-itself without advantages belongs to a kind of moral science fiction. They claim that moral behavior offers advantages that idealists obscure because they are unaware of them or do not wish to acknowledge them. Perhaps, that is why the Greek hedonist, Aristippus, in the fourth or fifth century B.C. proposed a concept of morality more credible to some people. He proclaimed enjoyment and physical pleasure as the highest of virtues. Throughout the millennia Aristippus has had no lack of followers. There will be people in the Age of the Genome who will heartily agree with Aristippus. They may look forward to the day when we shall identify the combination of genes that activate the neurotransmitters, serotonin, dopamine, and other euphoria-creating enzymes. Then these hedonists might gain lasting genetic "happiness" with less effort than it takes to obtain the common street drugs in use today. Nothing that previously existed could produce permanent "enjoyment" as efficiently as genetic engineering once the human genome's pleasure-producing characteristics are mapped and molecular biologist discover how to activate them.

The French philosopher, Henri Bergson, (1911), a Nobel Prize laureate, coined the term "creative evolution" to describe developments that occur independently in the human mind. Bergson's "creative evolution" allows for a touch of craziness within us that those who expect humans always to act rationally fail to take into account. Perhaps a recent view of our touch of craziness puts it in more acceptable terms. Researchers discovered that chaos exists normally in the human brain. Because it offers an alternative to "every man has his price," our touch of craziness leads us to our greatest innovation—behavior valued as a thing-in-itself. It enables us to adopt morality simply because we want to be moral persons. This isn't natural, logical, or rational, and that is why today morality still remains a subject of controversy and confusion. However, we should expect that in the history of life and the universe unprecedented developments can transcend the past and lead us into new directions. Because it offers an alternative to "every man has his price," our touch of craziness leads us to our greatest innovation—moral behavior that transcends certain limitations of our circumscribed biological nature. We shall discuss this further in the next chapter.

In summary: life on earth evolved about three billion years ago, and was an innovation in that life was able to replicate itself and to struggle to survive. Our human ability to make abstractions that seem to override the precise laws of biological nature represents an innovation of similar magnitude. In the future, wars, new diseases, and damage to our environment could cause us to become an endangered species. However, the greatest threat to our survival as human beings may come from our discovery of the biological secrets of life without fully understanding the meaning of morality.

In the next chapter I shall compare the writings of Nietzsche with the words of Jesus whose teachings illustrate what transcending could accomplish in the age we are now entering.

CHAPTER NINE

NIETZSCHE'S SUPERMEN AND THE PHENOMENON OF JESUS

The similarities and differences in how the 19th century German philosopher, Friedrich Wilhelm Nietzsche and the historical Jesus of Nazareth viewed human life will help us become aware of some of our options in the Age of the Genome. Separated by 2,000 years, Nietzsche and Jesus tried, each in their own way, to teach humankind important truths that they believed were hidden from the world. The sayings attributed to Jesus and the writings of Nietzsche, both present perceptions of human nature and ways to improve it. I shall briefly review certain aspects of the lives of Jesus and Nietzsche and some of the events that contributed to the formation of their ideas.

The task is not a simple one. Approximately 60,000 books were written about Jesus in the 19th century alone and each of them gave his life a somewhat different emphasis. The following descriptions are not intended to contradict views others may hold. In an historical presentation of Jesus, Oscar Cullmann, professor of Early Christianity at the Sorbonne in *Jesus and the Revolutionary*, (1970), cites Albert Schweitzer's book, *The Quest for the Historical Jesus,*

(1906), in which Schweitzer cautioned against a modernized, distorted view of Jesus.

In *Who Was Jesus?*, (1970), Colin Cross, a writer of history and biblical archeology, makes my point when he compares the uncertainty of what Jesus actually said with the doubt some scholars have about the authorship of the plays attributed to Shakespeare. Cross concludes: "the works of Shakespeare exist, whoever wrote them, and in the same way the parables and preaching of Jesus exist whatever their precise origin." Historians often face the dilemma that facts and philosophy are interdependent in that facts must be interpreted within the framework of the person's philosophy. Reflecting on this elsewhere, I wrote (1986) that, "science without philosophy can only offer techniques" and "philosophy without science can only produce conjectures."

E. J. Goodspeed, former Chairman of the New Testament Department at the University of Chicago, in *The Life of Jesus*, (1950), tried to reconstruct some events in the life of historical Jesus by using fragments of information and the logical consistency of available data. I borrow the following from some of his conclusions: Jesus lived in Nazareth and grew up in a large family, in a Jewish home typical of the times. Because of widespread poverty and the occupation of the land by Roman conquerors, there was great discontent throughout Israel. The Jewish religion as practiced then seemed unable to solve the peoples' many problems. This affected the young Jesus who was concerned about their unhappiness.

In his search for answers to the questions that troubled him, Jesus went to the banks of the Jordan where his cousin, John the Baptist, preached to large crowds. John baptized some of the people and Jesus was among them. After he emerged from the river's muddy water, Jesus experienced a transformation. He saw the world and people's problems in a new and more hopeful light. He wanted to share his insight with the people of Israel because he believed that it would enable them to override their seemingly insoluble problems. In effect, it required them to reverse their view of the logical order of things. He decided to dedicate his life to this mission.

"Call me rabbi," the Hebrew word for teacher, Jesus told those who asked him how they should address him. A teacher who believed that people could benefit from what he wished to teach them would not advocate the use of gene technology as a substitute for his instruction.

Jesus used parables to teach the people that they did not have to despair. He declared that the Kingdom of God had arrived and could turn their lives around. He promised that in the Kingdom of God the last—the downtrodden and suffering—would be the first, and the first—the proud who thought of themselves as superior—would be the last. If those who had "sinned" repented, they would be forgiven and could enter the Kingdom of God. According to some Christians, the Kingdom of God has meaning restricted to the phenomenon of Jesus. Others see it as compassion, concern, and love among people who accept the Kingdom as God's will for humankind.

Nietzsche was born in Prussia, Germany and lived from 1844 to 1900. His father, a Lutheran minister, died when Nietzsche was five. The young Nietzsche was raised by his mother, sister, grandmother, and two aunts. His early studies of the history of the Greeks, particularly the nobles in the Golden Age of Athens, greatly impressed him. These nobles lived for the moment and could accept victory or defeat with equal composure without traces of guilt. This may explain why throughout his life he kept two distinctions in mind—noble and slave, man and woman. Robert J. Ackermann, a professor of philosophy at the University of Massachusetts points out in *Nietzsche*, (1990), "In his [Nietzsche's] own view, decadence begins with the blurring of these distinctions."

Nietzsche elaborated on Darwin's thesis of biological evolution and believed the moral conduct of the Judeo-Christian ethic was a "slave morality." He claimed that this morality permitted the weak to limit the self-realization of the strong. Nietzsche viewed men who were self-confident, assertive, outstanding achievers, and were usually, therefore, highly visible, as *Übermenschen*—Supermen. They had reached a superior level of human development. Nothing could be further from Jesus' statement that, "the meek shall inherit the earth." The meek were practically invisible.

Christian morality created what Nietzsche called a "herd instinct" among these invisible people. Not much could be expected of them. In contrast, his *Übermenschen* stood out high above the crowd in their abilities in music, art, literature, philosophy, as well as warfare. Such men would construct their own laws of morality and take responsibility for them and therefore never experience guilt.

Kelly and Tallon, (1967), called Nietzsche's philosophy, "biological pragmatism." A biological pragmatist living in the Age of the Genome would have no reservations in using genetic engineering to redesign people who were permanently trapped by "slave mentalities" and set them free. A biological pragmatist might also recommend gene therapy for clergy who Nietzsche claimed used the Judeo-Christian values to give people guilt in order to hold them hostage to the church. Nietzsche's rejection of metaphysics and the social ideas of the 19th century caused him to view the whole moral world order of his time as an invention "against the emancipation of the man from the priest." If he could have looked ahead to the Age of the Genome, he would have viewed nothing as great a disaster as religious leaders having decision-making roles in something as important as the application of gene technology.

Nietzsche viewed himself as a scientist and a philosopher but actually he resembled a crusader most of all. When he wrote *The Antichrist*, (1888), translated in the *Portable Nietzsche*, (1976), by Kaufmann, he called Christianity "the greatest misfortune of mankind thus far." He also wrote, "In the whole New Testament only Pontius Pilate, the Roman governor commands respect." We have to ask ourselves, what becomes of morality when the Judeo-Christian tradition, the repository of morality in the Western world, is viewed as a misfortune?

To some extent one could question the intent of Nietzsche's works. Nietzsche's thoughts touched on many subjects. His ideas were close to thinkers of the Enlightenment, as well as to the romanticists, logical positivists, existentialists, atheists, and sociobiologists. Some reviewers of philosophical thought consider him one of the most provocative and influential thinkers of the 19th century. A number of these view him as a "moral philosopher" since he consistently attacked hypocrisy, indolence, and intellectual cowardice. After his death, his writings were distorted and used to support Fascist and antisemitic world movements to which Nietzsche, himself, would have objected. His ideas were misused by the Nazis for propaganda to promote the ideas of racial superiority of the Aryan race and justify the persecution of non-Aryan minorities. Kaufmann points out that Nietzsche made favorable references to the Old Testament and to the Jews of his time. In his writings, Kaufmann also notes in *The Portable Nietzsche*, (1976), that, "Any attempt to pigeonhole him [Nietzsche] is purblind."

Kaufmann made a surprising statement when he wrote, "many Christians feel they understand him [Nietzsche] best." How can one explain this in view of Nietzsche's bitter denunciation of Christian theology? Perhaps this was because Jesus believed that the rigid religious laws of Israel were out of step with human needs and Nietzsche thought the same of the religious theology of 19th century Europe. Jesus's declaration that "The sabbath was made for man, and not man for the sabbath," (Mark, 2:27), is at odds with the Roman Catholic Church with over two thousand canons to regulate every aspect of Christian life. The Church seemed to resemble a spiritual offspring of the Pharisees more than the spirit of Jesus who spoke out against the strict and literal adherence to religious laws. Had Jesus and Nietzsche lived in Spain in 1478, there is no doubt that both would have been called heretics and put to death by the Inquisition.

Even their sentiments regarding children were similar. Both thought of children as future actualizers of the potentialities of the universe. Mitchell, in *The Gospel According to Jesus*, (1991), described the large crowd who waited for Jesus' arrival in Judea after he left Galilee, including children who were brought to be blessed by him. The parents were rebuked by the disciples for having brought their children, but Jesus told them, "Let the children come to me and don't try to stop them; for the Kingdom of God belongs to such as these."

Nietzsche wrote in *Thus Spoke Zarathustra* (quoted by Kaufmann, 1967), "A child is innocence...a new beginning...a first movement, a sacred 'Yes.'"

Nietzsche's Supermen excelled in overcoming and can-do. Jesus would have had nothing against the idea of Supermen as far as their talents were concerned. However, he would have referred to them as "Superpersons" since throughout his ministry he respected women and their rights more than was usual 2,000 years ago in Israel. Jesus believed that *all* people were potentially *Übermenschen* and could rise to the highest level using the power of their own inner visions. He would have disapproved of the Supermen Nietzsche described because of their pride, feelings of superiority, and view of themselves as beyond sin and guilt.

The difference between the views of Nietzsche and Jesus will be critical to people living in the Age of the Genome. Jesus had little hope for the proud and self-assertive who considered themselves above good and evil. Jesus would have believed that the people described by Nietzsche as a "herd of sheep" could more easily gain entrance to the Kingdom of God than the Supermen admired by Nietzche.

Nietzsche's theme—the drive to power—could be seen as representing the flow of the dynamics of the universe. When driven by pride, it could lead to ever greater accomplishments, acquisitions, and gains. William J. O'Malley points out in *America*, (1994), that "Jesus did not consider wealth a sin..." His concern was not with wealth itself but rather the exploitation of others often associated with the acquisition and maintaining of wealth. Jesus taught that the flow of the dynamics of the universe should be directed to compassion and service without ulterior motives. As mentioned in chapter 8, the 18th century philosopher, Immanuel Kant's perception of morality echoed this. In spite of 2,000 years of exposure to the teachings of Jesus and the Eastern sages that condemned pride and acquisitiveness, some people continue to act as if they were disciples of Nietzsche. This occurs to me every time I hear the stirring and triumphant words and music of "Onward Christian Soldiers, marching as to war, with the cross of Jesus going on before"— so like Nietzsche and unlike Jesus.

In regard to the meaning of life, Jesus and Nietzsche differed tremendously. For Nietzsche, God was dead and a Kingdom of God had no meaning. Jesus advocated transcending life's problems peacefully. Nietzsche chose to overcome them by meeting them head-on with courage. Transcending offers humankind an alternative that requires even greater courage. Resisting the use of power to manipulate our genome may take the greatest courage of all.

The message that Jesus brought to the people could mean that a can-do world alone could not raise humankind to a level that offered life's deepest meaning. There is nothing wrong with can-do in itself and transcending was not intended to replace it. The life of historical Jesus was, in itself, a magnificent can-do. Nietzsche recognized this and included the name of Jesus on his list of Supermen.

Beyond can-do, people require transcending to bring out the gentler qualities they also have within themselves. They can only actualize the potentiality of the universe that would bring serenity into their lives by transcending the struggle discussed in chapter 6. Only by transcending can one reach the Kingdom of God and thereby actualize previously dormant potentialities of the universe that will bring good will into the world. Jesus did not hesitate to struggle with those who violated a moral principle as did those who exploited the sanctity of the temple. Forgiving must not be confused with overlooking.

O'Malley, (1994), believes that the clearest insight into the treatment of sinners by Jesus is found in the story of the Prodigal Son (Luke, 15:11-32). A son left home with money that his father gave him and soon frittered it away. Later, he realized his mistake and returned home with a confession on his lips. His father welcomed his return with a big party. His brother, who remained home, sulked at this elaborate welcome. The parable teaches that we need not be punitive. The celebration was not only for the son's return home, but also because he recognized his error. His father took the Big Picture view of the event and did not permit a fragment of reality—his son's lack of good judgment—to represent the whole of his son's reality. In genetic engineering the Big Picture view must be kept in mind. It is that of the universe and that of humankind and not confined to the view of the gene.

Another of Jesus' parables brings this out clearly. He likened the Kingdom of God to "a mustard seed, which is smaller than any other seed: but when it grows up it becomes the largest of the shrubs and puts forth large branches, so that the birds of the sky are able to make their nests in its shade" (Mark: 4:31). This parable suggests that transcending involves a change from the small picture view of the seed of self-interest to the Big Picture view of compassion and responsibility for all living creatures. The Big Picture view grows outward from within us and since we are only the small picture, we resemble the smallest of seeds. A seed within us contains the potentiality that gives rise to the Kingdom of God. It transcends the insignificance and limitations of our human nature to branch out and become "the largest of the shrubs." Transcending brings into play a relationship between the self and the world. Under those circumstances our genome may remain relatively safe from manipulations in the Age of the Genome. In his enthusiasm for his Supermen's ability to express their assertive feelings, Nietzsche failed to appreciate the significance of transcending in the way Jesus advocated.

Supermen or Supersalesmen may well be the ones who will make the major decisions on how gene technology will be applied in the Age of the Genome. Would they resist the temptation to use genetic engineering or genetic cloning to perpetuate themselves? In that case, meekness is certain to disappear as a human characteristic.

Nietzsche offered four directives to his readers: (1) don't remain like animals; (2) don't listen to the moralists; (3) decide right and wrong for yourself; and (4) seek eternity by creative efforts. He believed that the common people he called "a herd of sheep" would not be able to follow these instructions. It shocked his readers because he presented his views with forceful rhetoric at a time when the democratizing winds of the Enlightenment had swept over western Europe. Nietzsche's distinction between two categories of humans— the Supermen and the "herd of sheep"—caused many people concern. I have already mentioned that the Nazi and Fascist dictators used Nietzsche's classification of people as inferior or superior to justify their actions. This could not have occurred had they accepted Jesus' view of humankind. Communism's avowed objective, "From each according to his ability; to each according to his need" seemed compatible with teachings of Jesus. However, when this idealistic goal was enforced by a dictatorship, it lacked a requirement of morality—individual freedom of choice. The failure of Communism illustrates that the best intentions in the world if corrupted, obtain the worst results.

The Crucifixion of Jesus has been interpreted in various ways. Cullmann, described as "one of the world's distinguished New Testament scholars," writes in *Jesus And The Revolutionaries*, (1970), that it is "acknowledged by the majority of the scholars...that the legal condemnation was pronounced by the Romans rather than the Jews." He points out that the Romans were not interested in the religious squabbles and controversies of the Jews. They acted only on what they saw as a political threat to their rule. On the cross, as was customary, the Roman's posted their verdict, "The King of the Jews," which they judged as a political threat. Mitchell,

In The Gospel According To Jesus, (1991), sums it up with, "...it is clear that the Romans executed Jesus as a dangerous revolutionary."

Jesus died too young, Nietzsche lamented. He wrote in *Thus Spoke Zarathustra*, (1883-1885), translated by Kaufmann, (1976), that Jesus, "himself, would have recanted his teachings had he reached my age." Nietzsche in *The AntiChrist*, (1888), translated by Kaufmann, (1976), viewed Jesus' death as a "guilt sacrifice" and added that it was the "most barbaric... sacrifice of the guiltless for the sins of the guilty!" He called the glorification of the whole scenario of Jesus' death and resurrection a "ghastly paganism!"

A "Marginal Jew" is a term that Father Meir, a Catholic priest, used as a title for his book on Jesus, (1991). Meir described him as disgraced, flogged, and ridiculed while on the cross. Transcendence from there to the role of the "Son of God" provides a vision of truly heroic proportions. Nietzsche failed to see that the story of Jesus' death and Resurrection, whether it was on actuality or a legend, represents a concept of the near limitless power of transcendence.

The phenomenon of Jesus personifies the "good news" that almost anything is possible for humankind—*for humankind*—not merely for Supermen. Contrary to what Nietzsche viewed in Jesus' Resurrection, it fitted the parable of "the mustard seed," which is smaller than any other seed: but when it grows it becomes the largest of the shrubs and puts forth large branches." Nothing written by Nietzsche matched the actualizing of the potentialities by Jesus.

Yet, in a way that Nietzsche may not have intended, he was correct in thinking that something relating to the death of Jesus resembled "ghastly barbarism." It fits the description of Christendom using the Crucifixion to justify the hate of the Jewish people to whom Jesus, himself, belonged. There are ardent Christians who revere Jesus among those who have made a mockery of his mission by persecuting the people Jesus would have protected. The distortion of his message leading to antisemitism throughout the centuries symbolizes his ongoing Crucifixion. It was not till 1965 that the Roman Catholic Church formally exonerated the subsequent generations of Jews from their alleged responsibility for the death of Jesus. However, by 1965, the exoneration was too late. Antisemitism had already spread throughout the world and taken root which enable Christians as well non-Christians to use it to enhance their perception of themselves.

We are still too close to it to fully recognize the irony of Christian-fomented antisemitism. Two thousand years from now, if we should survive that long, students of "ancient history" who look back at us from the Big Picture view that time sometimes provides will shake their heads in disbelieve that Christians who accepted a Jew as the Son of their God could be antisemitic, or tolerate it, knowing that his mother, Mary, his brothers and sisters, his disciples, and the first missionary of the Christian religion, Paul, were Jews.

In Jesus' day there was the promise of the Kingdom of God for those who transcend hate and prejudice and replace them with love, acceptance, and compassion. We can only hope that in the Age of the Genome the fear of the misuse of genetic engineering will cause people to adopt the kind of morality that the philosopher, Immanuel Kant, described. Perhaps, we need this fear as well as the promise of a Kingdom of God on earth to put the teachings of Jesus into practice.

Records show that at the age of 45, Nietzsche became mentally ill. Let us imagine he suffered a disturbing hallucination in which he saw himself in the country of his birth standing on a rise that overlooked a collection of grim-looking prison buildings. In a cluster of trees outside the compound stood the crematorium. Nietzsche noticed a uniformed guard below in the prison yard look up and call out to him, "Nietzsche, we are doing this in your name." Horror gripped him and he turned away. Not far from where he stood, he saw a young man weeping, his eyes on the crematorium below. Nietzsche wondered who he might be. When he heard his despairing cry, "My God, my God, why have you forsaken me?" Nietzsche knew who he was.

In the Age of the Genome, Jesus may not, in anyone's fantasy, appear at Dachau to weep. Instead, he might roam the earth to mourn for an extinct humankind replaced by a new species of genetically-designed Supermen. But on the other hand, let's not totally rule out the possibility that Jesus might find that human beings had successfully transcended greed and were able to turn the other cheek to hate without resorting to genetic engineering. In that case we could picture Jesus saying happily to himself, "at last the Kingdom of God has arrived on earth. It's a bit late but it's doing well."

In the next chapter I shall discuss some issues raised in the preceding chapters including the use of gene technology to alter human nature in order to improve life and create a better world.

their lives to the enjoyment of science, philosophy, the arts, and recreation. Nevertheless, problems will continue in interpersonal relationships, coping with personality disorders, and probably, in dealing with crime and international strife. If, in the next century, we attempt to solve these problems by redesigning ourselves using advanced gene technology, we would lose our capacity to transcend, overcome, and circumvent problems. These abilities would become useless and atrophy in people who were genetic engineered to find nothing in the world objectionable.

As we approach the time when the human genome is mapped, sequenced, and catalogued, we must consider the wishes of both those who approve and those who disapprove of the application of gene technology. As I have indicated earlier, I do not anticipate much opposition to its use to replace or repair defective genes that cause diseases. On the other hand, genetic engineering used to improve food products has already encountered opposition and will continue to do so. Some nutritionists, environmentalists, and other concerned persons object to genetically altered food because they fear unanticipated side effects. A small group of people will continue to oppose any genetic "tinkering" with nature. Those who favor using gene technology to improve human life may think that too much caution would deprive the world of dependable sources of food. They are confident that improved methods of testing genetically altered food will prove some applications safe and others not safe. They predict that in time this will create confidence and that objections to using genetic technology to improve food will eventually cease.

Throughout the book my concerns have centered on the use of gene technology beyond food production and medical applications. In chapter 1, I mentioned that parents will have a great responsibility if they have the option to select the physical, mental, and emotional genetic make-up of their children. In spite of controls that will be established after the techniques are perfected, an illegal market for gene technology will become available. The lure of great wealth derived

CHAPTER TEN

CONCLUSIONS

In the past two years, chromosome cartographers increa
the number of genetic markers they located on the hur
chromosomes from 814 to over two thousand. These marl
help chart the relative position of the genes on our DNA.
estimated five thousand markers on the expanding hur
genome map (six thousand on the mouse map) will be i
tified within a year. Specially designed computers
automated data-handling techniques help speed up the w
As we increase the density of markers charted on the D
it will become easier to locate specific genes. Mole
biologists will complete the task of mapping the entire hu
genome before the end of the next century. Therefore, w
appropriately call the 21st century the "Age of the Geno

In the Age of the Genome, we will rely on nonpoll
sources of energy and produce sufficient food without th
of insecticides. Genetic engineering will prevent many i
ited diseases and provide gene therapy for patients v
prognoses would be hopeless without it. New technol
will harness the power of sub-atomic particles
bioengineered bacteria and viruses will do much of the
presently performed by human labor. It could bec
Golden Age in which people would have the leisure to (

from providing genetic engineering will be irresistible to some entrepreneurs. We are only speculating because it is still questionable whether bioscientists will ever be able to create significant personality changes using gene technology. In the future we can be sure that some can-do descendants of *Homo habilis* will take up the challenge. They will attempt to redesign human beings regardless of laws and religious prohibitions. These entrepreneurs might resemble Nietzsche's Supermen described in the preceding chapter as "above guilt and evil."

Less ambitious programs that involve identification of a person's genetic characteristics will be in full swing long before the entire human genome is mapped. In the Age of the Genome, genetic profiles will become available to employers, insurance companies, prospective mates, and others interested in a person's hereditary make-up. Then we shall add a new elitism based on genetic profiles to the long list of ways some people use to feel superior to others.

Studies continue to suggest that some violent-prone individuals may have a genetic predisposition to an arousal of intense hostile emotions. When it becomes possible to do so, social pragmatists may propose that genetic alterations be used to eliminate violent and antisocial behaviors. Their goal will be to help create a non-violent world, a Kingdom of God on earth, by means of genetic engineering. They will ask: "Would enhancement of human life with gene technology be morally wrong when people, themselves, request it?" Prozac, imipramine, and other neuroleptics create biochemical changes in the brain. They would consider that changing human behavior by means of gene technology might be preferable to using mind-altering medications.

I do not believe that genetic engineering should be used for any reason other than for physical health or to prevent mental illness. In the pages that follow I will expand on some essential points that contribute to my decision.

In contrast to genetic engineering, biochemical personality changes created by medically prescribed drugs can be reversed. They require cognitive participation by the patient, even if only to the extent of taking a pill. This act, though a minor one, nevertheless represents the essential element of self-involvement and control. Freedom of choice would not be passed on to the future generations who would inherit the genetic make-up of their parents. The title of an early book on genetic engineering by June Goodfield, *Playing God*, (1977), reflects my concern. "Playing God" is an appropriate description for those who would like to redesign people. A play-god can produce only a play-kingdom of god and such a play-kingdom only requires a play-morality. We find such kingdoms headed by play-gods existing today as cults. Play-gods cease playing while they strip their followers of their judgment, possession, and wealth. We must avoid play-gods and create a better world—a Kingdom of God—the hard way, using our own best efforts to overcome obstacles.

Many socially-concerned people doubt that we can establish a moral world before the mapping of the human genome is completed. Since there are no alternatives, we must not give up before we start. Surely, a species able to conceive of the idea of a commitment to morality can encourage more of its youths to live responsibly. Perhaps, if we portrayed a commitment to morality as an exciting challenge we might be more successful.

In the Age of the Genome we shall continue to need rewards and punishment to make some people comply with the rules of social behavior. These people when conforming would, of course, be hypocrites. Hypocrites have no value in churches but society still needs them. La Rochefoucauld aptly said, "hypocrisy is homage, that vice offers to virtue." We need both, some people's "homage" to morality and the genuine morality of others to keep the human genome from exploitation.

In the previous chapter quotations from Jesus and Nietzsche show their faith in the open-mindedness children have to new ideas. Parents, teachers, the clergy, youth leaders, the entire community, as well as all forms of the media, must begin now to make today's children aware of the unusual and intriguing characteristics of morality. We should encourage our youths to replace their highly visible "heroes" carrying assault weapons with exciting heroes who struggle to adopt moral values. If given a choice, many of our young people would find values that center on human concerns more appealing than those that consist of leverage—if you don't do what I want, I'll do what you don't want. Role models having these selfish values are now shown on our television screens and may be seen swaggering on the streets of our cities. We can only hope for a moral world in the Age of the Genome if we start now to give our children's lives a new sense of direction.

I shall summarize some points made in the book. In chapter 2, I presented the view that all reality remains eternal. Regretting acts committed or words spoken take on new meanings when they are perceived as changing the universe itself. There is more than appears at first glance to Jesus' exhortation: "repent!" From the definition of reality given in this chapter, everything that happens has significance. When something comes into being, it continues to exit as an eternal attribute of the universe.

Actualizing the potentialities of the universe, I said in chapter 3, provides a sense of adventure. Everyday things we take for granted are in themselves miracles of actualization. Obtaining pleasure from "things *not* of the spirit" causes some people to feel guilty. They should remember how much the empty Coca-Cola bottle impressed the fictitious motion picture Bushman who assumed it was a work of the gods. All things that reflect human ingenuity actualize potentialities of the universe and using them is fun—not sin! Confused moralists fail to understand that enjoying clever modern inventions does not represent succumbing to "material values." They are actualized potentialities in common with pyramids and cathedrals.

In contrast, some of the ideas originated in the Age of Enlightenment are described in chapter 4. Life was viewed as something that must be demonstrated before it could be accepted as real. Manifestation became the fragment of reality that in the eyes of many was the whole of reality. Long before this, the human penchant for manifestation had been recognized as a problem. Even in ancient times, vanity was decried as a cause of human downfall. However, manifestation as a goal in life increased in the 17th century during the Age of Enlightenment. It became almost an obsession from which we still suffer today as we enter the Age of the Genome. This obsession leads us to view ourselves as the source of our own souls. As mentioned in chapter 5, some neurobiologist ascribe the origins of phenomena to our internal mental equipment and bypass the mystique inherent in the potentialities of the universe. A fragment of reality is seen by them as all of reality—a dangerous misperception discussed in chapter 2. I maintain that our potentialities do not originate within our human skulls. It's the other way around. The content of our skull represents these potentialities.

The "Gospel According to John," (John, 1:1) starts with, "In the beginning was the Word, and the word was God . . . " Throughout the book I have called the *Word*, "potentiality." Actualizing the potentiality of the universe can also be seen as "doing God's work" which Bishop Tutu says is the meaning of life. This is consistent with discovering the meaning of life in living it. Our life has meaning because everything we do and think contributes to eternal reality. The laws of nature do not allow the existence of nothingness which some equate with death. The idea is important since the perceptions of life and death will play a significant role in the future application of gene technology.

In time, we shall be able to control the genes that program the length of life of various species. Therefore, Hamlet's question, "To be or not to be?" will gain new relevance in the Age of the Genome. Eukariotes—plants and animals—have genes that turn their lives off at a predetermined time. These genes have been called, "death genes." It is correct to say the days of our lives are numbered, not only by circumstances, but more surely by the genetic clock within our "death genes." Altering the timing of these genes or turning them off could, theoretically, extend our lives or even give us immortality. Immortality as a theoretical possibility of gene technology leads us to take a closer look at the fear of death—a subject pondered over by artists, poets, philosophers, and Freudian psychoanalysts.

Traditionally, we view the fear of death as dread of the unknown. In some cases this may be true but, as I mentioned in Chapter 6, the idea that death forces us to cease actualizing the potentialities of the universe causes anxiety. Most religions describe some form of afterlife where actualizing continues. In ancient times, warriors were often buried with their weapons and craftsmen with their tools so that they could resume their work in their afterlife. Some rulers were buried with riches, food, and symbols of their power as well as their wives and members of their court, so that they could continue to rule in the next world. "We shall not cease to exist" is expressed by countless monuments erected all over the world. They are echoed by ancient mummies waiting patiently to be awakened. They failed to view their thoughts and acts in life as eternal monuments of the universe.

Birth is eventful because every new life represents another actualization that keeps perpetuating a species. Therein lies the argument against engineering ourselves for immortality. Obviously, if death ends so must birth. Birth is a weapon in the battle that gives the universe it's best chance for actualizing its potentialities. It is a two-edged sword since it can lead to overpopulation. Its mechanism is explained in an article, "Tricks to make DNA beget DNA" in *Science News*, (1994).

> Living systems expand exponentially: two DNA strands beget four, which beget eight, then 16, then 32, and so on. Chemical systems increase incrementally, from one to two to three copies and so on.

Overpopulation could eventually destroy our chances to live peacefully in the space-limited Kingdom of God. "Be fruitful and multiply!" was a command given when the population of the world was less than that of any one of the world's larger cities today. Were God still on speaking terms with human-kind He would, at this point say to us, "Enough!"

Death raises serious ethical questions. There is a legitimate concern with ending life for those suffering from terminal illnesses. Our definition of morality in chapter 8 makes it clear that moral people must respect individuals' decisions that pertain to their own lives. Bronowski's definition in chapter 8 suggests that it is immoral to prolong persons' lives with medical devices against their will. Today, dying patients are victims of play-gods. When the time arrives to die, only play-gods who do not understand the meaning of morality would attempt to deprive terminally ill persons of their right to die with dignity.

I discussed the role of struggle in the development of the human essence in chapter 6. It made us a resilient, persevering, assertive and, at the same time, a dangerous species. The qualities that distinguish us and make us unique get us into trouble. Struggle doesn't have to stomp around in hobnail boots. Often it accomplishes more when it walks softly in slippers. The contradictions inherent in our nature cause us to yearn for harmony as we struggle to remain human. Our *raison d' être* would end if humans succeeded in establishing only harmony. If we found our longed-for harmony in a Kingdom of God, it would be planted with trees bearing forbidden fruit to tempt us. Temptation is the origin of the struggles that helps to keep us human. Jasper, in *The Way to Wisdom*, (1954), reminds us that we must even be critical of "the love of wisdom that does not struggle with itself."

I try to identify some priorities in chapter 7. Not too surprisingly, I placed the survival of our species first. I did not do this to please our "selfish genes," but because our species has the greatest ability for actualizing the potentialities of the universe. I suggested, in chapter 8, that a commitment to morality gives the universe feedback which changes its configuration. Actualizing a potentiality of the universe is a reciprocal process, circular, not linear. It has some of the elements of chaos that add a measure of unpredictability to an outcome. That applies to the effect morality will have in the Age of the Genome. Conover, (1962), points out that morality "is our second nature rather than our first nature." He recognized that "human beings do not become moral *without* [italics are Conover's] training and education." He believes that moral training must begin "from the earliest days of infancy . . . neglect of moral education will result in increase in disorder and evil." These were prophetic words for the following decades in which we have witnessed the decline of moral education in our entire society.

Perhaps, a wider acceptance of the Kantian view of morality as defined in chapter 8 has had to await the mapping of the human genome. More than any other scientific achievement, the fully mapped human genome will highlight the oneness of humankind. Regardless of the forms of its expression, at the heart of genuine morality lies the recognition of the relatedness of all the world's peoples. This is important since those who limit their moral commitments to selected groups deprive morality of significance.

In defining and describing its characteristics, the idea of a commitment to morality becomes complicated. A simpler view of morality has long been recognized and fully validated by experience. Morality creates itself without effort or philosophical contortions within the hearts of those who love with understanding and acceptance.

Love begets a morality that meets all of the requirements of Kant and Bergson. If one loves another person, loves humankind, or falls in love with the Big Picture view of life, one becomes a moral person even if one doesn't know what the word, morality, means. This is because the most accurate definition of morality is really the simplest one: morality is the behavioral expression of love.

To paraphrase the words of Conover, love transcends our "first nature" and reveals our "second nature." Love makes us willing, at times, to meet the needs of another person at the expense of our own needs. Love converts sex into sex-with-love. Love is dedication and sacrifice—the enablers of transcendence. Clearly, this is the "good news" Jesus brought when he said, "love your enemy." Those who love their enemy will love all else.

In chapter 9 in his *Will to Power*, Nietzsche put love into a different context (Kaufmann, (1976)).

> ...the love of power is the demon of man. One may give them everything—health, nourishment, quarters, they remain unhappy for the demon insists on being satisfied. One may take away everything from them and satisfy this demon: then they are almost happy.

Nietzsche was right about the "demon" who must be satisfied. In terms of this book, the "demon" is our drive to actualize the potentialities of the universe. Since the demon may become fierce if deprived of the chance to actualize, we shall be safe only if every person has an opportunity to play a role in life that contributes to society. Then the demon may be tamed to work constructively for us.

Here in the Western world where gene technology received it start, Christendom has had an overriding influence. It gave us inspiring visions of transcendence by transforming a crown of thorns into a halo. However, in some respects, we must agree with Nietzsche that Christianity must reverse the illogical order of its demands. It must "repent" locking Jesus into heaven, out of earth's way, where priests and ministers can safely put words into his mouth that would have made him gag. The churches can list their good intentions, their charities, count their martyrs, and roll out scrolls of good deeds. They provide their church members with a needed sense of community. They enable them to obtain glimpses of the Big Picture that inspires reverence and awe. However, this is not enough. Churches must shed self-serving militancy in condemning those who differ from them. Instead they should become *activists* not as some do now, against people's right to differ, but as forces against racism, prejudice, chauvinism, and irresponsibility. Action taken against hate is a prayer magnified a thousand times. In James, (1:22) we find, "Be doers of the Word, and not hearers only, deceiving yourselves." The phenomenon of Jesus shows this meant "doers" that bring about acceptance and end discrimination.

Almost all religions, in one way or another, have social concerns. They tend to the sick, assist the disabled, collect money for the poor, care for widows and orphans, and make announcements that reflect concern for humankind. This is all to the good but it is insufficient to actualize the new potentiality of the universe inherent in transcendence. To meet the needs of the Age of the Genome, religious leaders of all denominations and persuasions must take their cue from the Big Picture view ascribed to Jesus by H. James Birx in *Man's Place in the Universe*, (1977). He writes, Jesus "is no longer a historical figure. Instead . . . he takes on a cosmic and dynamic dimension." The antagonism the world's religions have toward each other, their concern with self-perpetuation places them into a category with Dawkins' *The Selfish Gene*, (1989), instead of within the framework of the Sermon on the Mount.

Fortunately, we shall have a powerful new ally in the near future—not in the role of a Messiah from Heaven—but surprisingly, one Nietzsche so aptly called the "Demon within us" on earth. It is the drive to power that Supermen would use to control the world by means of gene technology. Fear that they may then convert the rest of humankind to a "herd of sheep" may be what we have needed all along. Unfortunately, love by itself has not changed the world. Love combined with fear may have a better chance of doing so. I use "fear" in the sense of "The fear of the LORD is the beginning of wisdom." (Psalm 111:10). That wisdom includes awe, reverence, and respect for life as it *now* exists. It may give us the final spurt of motivation that will impel us to adopt a morality that will save humankind in the Age of the Genome. Let us never forget that we have the ability to overcome because we are a can-do, can-transcend species genetically engineered that way by nature aided by the values we adopt.

Throughout the book I use "universe" as some readers within a different context might say, "God." As a scientist I wished to remain as much as possible within the realm of science. The words in this book may be placed into any context acceptable to the reader. At times, I have anthropomorphized and described the universe in a way that makes it appear alive. I shall borrow a phrase from William James, the American philosopher and psychologist, which explains why I felt I needed to do so.

In *The Varieties of Religious Experience*, (1958), James wrote, "the universe is not a mere It to us but a Thou, forced on us we know not whence . . . " This is the only perception of the universe that can reveal the Big Picture view of mapping the human genome within the context of Reality, Morality and Deity.

EXTENDED GLOSSARY

* indicates the word has special or an extended meaning in terms of this book.

animal-human differences - the structure of our human foot associated with our upright gait, the relative size and complexity of our brain that leads to can-do capabilities, the long period of human infancy and early childhood dependency that promote nurturing and learning, the capacity for language and making abstractions that reveal unactualized potentialities of the universe.

antibiotic - one of a group of organic compounds that can destroy bacteria and inhibit the growth of microorganisms. However, an increasing number of strains of bacteria are becoming immune to these compounds by mutating. This poses a threat to human life and a major medical challenge to genetic technology.

antisemitism - a dislike of Jews by those who generalize and distort in order to blame them, as an ethnic or religious group, for wrongs for which Jews are not responsible. Sometimes antisemitism is rooted in an emotional inability to apply Jesus' teaching of tolerance and love. Antisemitism rejects Jesus' own ethnicity. Symbolically, it represents a second crucifixion of Jesus.

bacteria - one-celled microorganisms which multiply by simple division. From the point of view of adapting to their own and other environments, they are among the most successful life forms. Some are useful to humans and others will be even more valuable when they are bioengineered to serve human needs. (See antibiotic)

Big Bang - the primeval explosion that scientists believe started the universe an estimated 15 or so billion years ago. The theory has been debated for different reasons and so has the implication of the name given it. However, the term is still in common use among cosmologists to signify the beginning of our present universe.

biotechnical - the application of technology to biological processes for industrial, agricultural, and medical uses.

* can-do - may be viewed as the human effort to give a directional focus to the dynamics of the universe with the intention to overcome obstacles. Can-do results in changing the conditions that hinder the improvement of human life. Transcending overrides the self-centered aspects of can-do. (See transcend)

chaos - the occurrence of an event that causes a succeeding series of events in ways that do not follow the rules of traditional statistics. It appears as random, erratic, and unpredictable. A new science has evolved from the study of chaos that uses a nonlinear approach to probe for order and patterns in chaotic events.

chromosomes - one of a group of thread-like bundles of different lengths and shapes that contain the genes that make up the genome of a species. Their number varies among different species. Chromosomes usually occur in pairs. Most cells in the human body normally contain 23 pair giving our genome a total of 46 chromosomes that, together, hold an estimated 100,000 or more genes.

deoxyribonucleic acid - See DNA.

DNA - initials that stand for deoxyribonucleic acid. Mainly found in chromosomes that contain the hereditary information of organisms.

* dynamics of the universe - forces rooted in the nature of the universe that permeate everything and account for the "push" that causes things to happen. Life gives them a directional focus and human life can redirect the focus by transcending. (See transcend)

enzyme - a compound that serves as a catalyst in biochemical reactions. Enzymes are complex proteins that create change in the function of other substances without changing themselves.

eugenics - a theory which proposed that the human race would be improved by restricting propagation among those considered genetically unfit. It aimed to encourage breeding between individuals with genetic traits considered desirable as intelligence, health, and leadership. The movement was associated with a sense of white, Anglo-Saxon superiority. It could become a forerunner of genetically redesigning humans in the Age of the Genome.

eukaryote - an organism whose DNA is enclosed by membranes to form a nucleus within a cell. In contrast, the DNA of prokaryotes lies free in the cytoplasm. Eukaryotes are characteristics of plants and animals; prokaryotes evolved first and comprise mainly bacteria.

existentialism - a philosophical movement that developed in Western Europe in the 20th century. It maintains that there is no fixed human nature. Humans must have free expression rather than conform to given rules. Its emphasis is placed on responsibility for one's own actions. Existentialists point out that this could lead to feelings of helplessness and despair.

gene - the basic unit of hereditary material located on a chromosome that, by itself or with other genes, determines a characteristic of an organism.

gene, dominant - carries a trait in a person's genetic make-up that is expressed in the life of the individual. (See gene, recessive)

gene, recessive - carries a trait that is not expressed in an individual's life unless paired with another recessive gene carrying the same characteristic. It is passed on to offsprings who will not exhibit its trait if it is paired with a dominant gene that lacks it. Recessive genes are more apt to carry undesirable characteristics including genetically caused diseases

gene therapy - consists of repairing or replacing damaged, faulty, genes that cause disease. Ethical questions arise about its application to prevent personality disorders or to correct genetic tendencies considered disruptive to society.

genetic engineering - is also called, recombinant DNA technology. It involves the direct introduction of foreign genes into an organism's genetic material by micromanipulation at the cell level. It enables correcting defective genes as well as gene transfers between widely different forms of life and may change the inherent nature of an organism. Without a framework of reality, morality, and deity, its application to human life could lead to the extinction of our present species and replace it with one having different characteristics.

genome - a complete set of chromosomes carried by each cell of an organism. (See chromosome) The human genome contains all of the genetic characteristics that individuals will develop over their lifetimes. Mapping a genome enables geneticists to alter different forms of life including our own.

* happenings - actualized potentialities of the universe.

* harmony - a theoretical state of tranquility, balance, and quiescence often longed for. If we attained it, we might not be able to tolerate it without losing our unique human characteristics. Struggle is present whenever there is challenge. Struggle needs not involve violence, or strife. It will occur naturally in the minds of people who see both sides of any issue. We may experience a feeling of harmony by accepting struggle as an aspect of life and making peace with it.

heliocentric earth - the idea that the earth is the center of the universe and everything spins around it.

Homo habilis - can be translated as "the handyman" or as I have used it, our "can-do" ancestor. Some dating places this group at having lived from 1.2 to 1.8 million years ago. Their remains are found with crude stone tools fashioned by the use of other modified tools. (see tool) Earlier crude stone tools, estimated as 2.5 million years old, have also been found in Northern Africa.

hormone - a chemical messenger released by a certain type of gland and transported in the blood to a specific organ. It acts to stimulate growth, metabolism, sexual reproduction, and other body processes.

* human nature - there are two or more kinds of human nature. The book refers to the concept of a first and a second one. This distinction involves more than different kinds of behaviors originating in upper and lower brain centers. The first human nature is shaped by culture as well as by a person's genetic make-up. The second human nature results from transcending the first, enabling a commitment to morality.

* imagination - the most astonishing of all human actualizations of the universe. Within its framework it provides unlimited power to those who utilize it well. It serves to achieve transcending when used as "imaging"—a psychological technique for directing and vivifying the imagination to create a change of view and feeling. Imaging an enemy as a friend can transform hate into love. It is a requirement for the Age of the Genome. (See transcend)

* Jesus, (as a historical figure) taught that people have within themselves the power to transcend greed, selfishness, and avarice. His teaching made them aware that they had the ability to reach a state of life that would lead to a "Kingdom of God" on earth. It would bring them good will, happiness, and satisfactions. Following the teachings of Jesus would enable people to live together in peace as members of a caring community. The phenomenon of Jesus provides us with a role model for dedication. If more people would accept Jesus as their role model (instead of merely as a statue before which one kneels) there would be less reason for concern that gene technology would be exploited or misused in the next century.

* **Kingdom of God** - on earth (from the author's point of view) will arrive through transcending a concern for self and replacing it with concern for *all* of humankind. The use of the word "God" implies the Big Picture view which must be taken to achieve it. Sometimes, people who have suffered are less likely to be indifferent to the distress of others. Their concern for others enables them to enter the Kingdom of God more readily than those who lack compassion. The Kingdom of God embraces all living things as well as the environment. Nevertheless, it is species-oriented. The Kingdom of God for humans will not be a Kingdom of God for microorganisms dangerous to humans.

* **logical order** - used with "reversal of." It creates a distortion caused by taking events or happenings out of their logical sequence, their context, or out of place in a priority of values. Although such a concept remains subjective, a logical order of things is based on rational priorities. That is why a reversal of the logical order of things often turns into defeating one's purpose as when one puts the cart before the horse.

* **manifesting** - to show, or show off what one actualized. In this book, I have contrasted manifesting with actualizing the potentialities of the universe. On the positive side, manifesting occurs in communicating, teaching, sharing, and informing. Neutrally, it may be used to gain acceptance and reassurance. On the negative side, it's purpose is to call attention to one's accomplishments for vanity or because one desires additional status and recognition. I surprise myself and am appalled when, occasionally, I stop to ask, "are my intention at this moment to actualize dormant potentialities of the universe for the common good or to manifest those I have already actualized primarily to gain self-satisfaction? If oth-

ers tried this brief experiment their answers often, like mine, may become a source of New Years' resolutions.

* **meaning in life** - is found in living it. From a Big Picture view the universe may be seen as composed of all the happenings that occurred and will occur. Everything we do or think, helps write its history. For that reason we cannot avoid giving life meaning by whatever we do or even don't do. Those who actualize the universe's dormant potentialities *selectively* give their lives a special meaning.

* **morality,** *false or pseudo.* "Morality" that results from genetic programming, indoctrination, fear, or serves as a means to an end. We should not confuse indoctrinating with teaching or role-modeling. These can lead to the acceptance of a genuine morality.

* **morality,** *genuine* - results from a voluntary commitment to an ethical way of life in which there is consideration for others. In such a life one usually finds value placed on honesty, fairness, self-control, duty, and dedication. Morality's overriding criterion is that it is freely chosen without ulterior motives. Contrast this with today's popular goals in most human interactions solely based on, "what's in it for me?"

neocortex - the part of the thin, gray outer layer of the brain's cortex usually associated with human thought and higher intelligence.

neuroanatomy - the anatomy that deals with the nervous system and the brain.

nucleotides - comprise the mosaic that helps determine the functions of the genes. Nucleotides may be viewed as the building blocks of the DNA and RNA.

nucleus - an organelle of plant or animal cells containing the genetic information and controlling the cell's activities. Organelles are subcellular structure with a particular function. The largest organelle is the nucleus. Organelles allow division of labor within the cell.

* parts, more than sum of, - has sometimes been called, "beyondism." Usually the people who recognize it, view human life as a fragment of a greater reality. Humans are among the most unusual fragments that exist in that some of us continually search for our relationship to the whole.

protein - One of a large number of substances that are important in the structure and function of all living organisms.

* reality - All happenings and everything that is or was actualized since the beginning of the universe contribute to creating reality. Ideas expressed in words may actualize some potentiality more effectively than physical events as Bulwer-Lytton (1803-73), an English novelist, observed in *Richelieu* (1838), "the pen is mightier than the sword." Nevertheless, words and deeds, facts and fiction represent different kinds of reality and should not be confused or replaced by each other. Though both have afterlives within reality, one might say, each lives in its own kind of heaven. Science and pseudoscience are drawn from different sources and failure to differentiate between them leads to a confusion of all of reality and confounds can-do and overcoming.

recombinant DNA - DNA formed by crossing over genetic material from one form of life to another. It is sometimes used as another term for genetic engineering.

* sex - Sex involves the flow of the dynamism of the universe through living organisms that actualize potentialities of the universe non-living matter could not. The sex instinct pushes life to actualize reproductive potentialities whereas humans can devise additional or alternate ways of creating reality. Taking a Big Picture view of sex enables us to consider sex-with-love as a more advanced contribution to the nature of the universe than sex as simply a reproductive response to instinctual drives. This is because sex-with-love actualizes two potentialities of the universe simultaneously and leads to further actualizations for example, poetry and song.

* slave mentality I refer to people who are greatly influenced by the media as having slave mentalities or, as Nietzsche might have put it, the minds of a "herd of sheep." I consider them as the most vulnerable to all kinds of propaganda and, therefore, most likely future victims of gene technology.

Social Darwinism - a view of society that accepts the fundamentals of biological evolution as, for example, competition and survival of the fittest. Social Darwinists believe that competition brings out the best in people and without legal interference would create a moral society. Social Darwinists consider themselves as realists and oppose what they view as impractical idealism that would, in their opinion, lead to social deterioration and threaten the quality of life.

* **spiritual** - Average-sized dictionaries give up to sixteen different definitions for the word, *spiritual*. In some respects it has become a buzzword. Therefore, its meaning in a specific context is often unclear. Most definitions describe *spiritual* as "things of the spirit," those things whose awe-inspiring qualities lift them above material things. Though humankind represents only a fragments of reality, humans search for their relationship to total reality. This evokes feelings of reverence and might be an illustration of something appropriately called, "spiritual."

struggle - See harmony.

tool - a device that facilitates can-do. Apes and some other animals use tools occasionally but the later *Homo habilis* went one long step further. He and she discovered that an object they fashioned into a tool could, in turn, be used to modify other objects that became more complex tools. This was a breakthrough of great significance to later human development.

* **transcend** - The word, *transcend* means to go above or beyond something else and in so doing results in a profound change of view. To some, the word is used to describe a miraculous transfiguration while others see it as a result of insights that put important aspects of life into different perspectives. In religion *transcend* sometimes is used as being in touch with God. In meditation it could mean arriving at an

altered form of consciousness. Transcendence usually causes a change in emotions and motivation as when the focus of one's life turns from self to others. It may reveal a Big Picture view of an event previously seen only with a personal involvement. A commitment to morality involves transcending an exclusive focus on self-centered goals. I see it as the basis of a future form of psychotherapy appropriate for the Age of The Genome.

transgenic - pertains to organisms that have received genes from different organisms.

* **universe** - used in this book as composed of everything that happened in the span of its existence. Beyond that it also consists of the dormant potentialities that have not yet, and may never, become actualized. The dynamic flow that runs through all things "pushes" them to actualize the potentialities of the universe. More than anything else on earth humans respond to that "push." This greatly complicates our lives by constantly forcing us to make decisions.

* **visibility** - seeking visibility as used in the book represents an emphasis on manifesting over actualizing the potentialities of the universe. (See manifesting) The historical emphasis on visibility could stem from misunderstanding the nature of the universe leading to a false perception of the meaning of human life. Concern over a lack of recognition, response to indoctrination, or the influence of role models could account for an undue emphasis on visibility. This tends to drive other considerations out of one's mind and causes people to reverse the logical order of things. (See logical order)

BIBLIOGRAPHY

Ackermann, R. J. (1990) *Nietzsche—A Frenzied Look.* Amherst: University of Massachusetts Press.

Barash, D. P. (1977) *Sociology and Behavior.* New York: Elsevier.

Bellamy, E. (1889) *Looking Backward 2000-1888.* Boston: Houghton.

Bergson, H. (1911) *Creative Evolution.* New York: Holt.

Birx, H. J. (1977) *Man's Place in the Universe.* Arcade, New York: Tri-County Publications, Inc.

Bronowski, J. (1965) *Science and Human Values.* New York: Harper & Row.

Cohen, M. T., M.D. (1994) Quoted by Paul Rayburn in *San Diego Union-Tribune* (February 20).

Conover, C. E. (1962) *Moral Education in Family, School, and Church.* Philadelphia: The Westminster Press.

Corballis, M. C. (1993) *The Lopsided Ape—Evolution of the Generative Mind.* New York: Oxford University Press.

Crick, F. H. C. (1994) *The Astonishing Hypothesis: The Scientific Search for the Soul.* New York: Scribner.

_____ (1988) *What Mad Pursuit: A Personal View of Scientific Discovery.* New York: Basic Books.

Crick, F. H. C. and J. D. Watson (1953) "The Double Helix." *Nature.* April 25.

Cross, C. (1970) *Who Was Jesus?* New York: Atheneum.

Cullmann, O. (1970) *Jesus and the Revolutionaries.* Translated from the German by Gareth Putnam. New York: Harper & Row.

Darwin, C. (1859) *On The Origin of Species.* London: J. Murray.

_____ (1871) *The Descent of Man.* London: J. Murray.

Dawkins, R. (1989) *The Selfish Gene.* New York: Oxford University Press.

Diderot, D. (1754) *Pensée sur l'interpretation de la nature.*

Drummond, H. (1894) *The Ascent of Man.* New York: A. L. Burt.

du Noüy, L. (1949) *Human Destiny.* New York: New American Library.

Eliot, T. S. (1935) *Murder in the Cathedral.* San Diego: Harcourt Brace.

Frye, C. (1979) *A Yard of Sun.* Oxford: Oxford University Press.

Galton, Sir F. (1869) *Hereditary Genius.*

Goodfield, J. (1977) *Playing God—Genetic Engineering and the Manipulation of Life.* New York: Random House.

Goodspeed, E. J. (1950) *The Life of Jesus.* New York: Harper & Brothers.

Greunbaum, A. (1987) "Time, Irreversible Processes, and the Physical Status of Becoming." In *The World of Physics*. ed. J. H. Weaver. New York: Simon and Schuster.

Halacy, D. S. (1974) *Genetic Revolution: Shaping Life for Tomorrow*. New York: Harper and Row.

Harrison, E. (1985) *Masks of the Universe*. New York: Macmillan.

Hawking, S. W. (1988) *A Brief History of Time*. New York: Bantam.

Hopfield, J. J. (1994) "An Envisioning of Consciousness." Review of *The Astonishing Hypothesis*, (1994), by F. H. C. Crick. New York: Scribner. *Science*. 263: 696.

Huxley, A. (1932) *Brave New World*. New York: Harper Collins.

James, W. (1958) *The Varieties of Religious Experience*. New York: Mentor Books/New American Library of World Literature.

Jasper, K. (1954) *The Way to Wisdom: An Introduction to Philosophy*. New Haven: Yale University Press.

Kauffman, S. A. (1993) *The Origins of Order*. New York: Oxford University Press.

Kaufmann, W., ed. (1976) *The Portable Nietzsche*. New York: The Viking Penguin (12th printing).

Keddie, N. R. (1993) Postscript: "Comparative Reflections on Hindu Extremism." In *Resurgent Hindu Fundamentalism* by Stanley Wolpert. In *Contention*. Vol 2. No.3, Spring. Bloomington: Indiana University Press.

Kelly, W. L. and A. Tallon (1967) *Readings In The Philosophy of Man*. New York: McGraw-Hill.

Kent, S., ed. (1989) *Farmers as Hunters - The Implications of Sedentism*. Cambridge: Cambridge University Press.

Kent, T. C. (1986) *Conflict Resolution: A Study of Applied Psychophilosophy.* Woodbridge, CT: Ox Bow.

—— (1989) *Speculating on Roots of Human Behavior Beyond Biological Origins: Pragmatic Reasons for Not Doing So—Possible Consequences of Doing So.* Paper presented at annual meeting of the Pacific Division of the American Association of Science, 11-15 June, at California State University, Chico, California.

Lee, T. F. (1991) *The Human Genome Project—Cracking The Genetic Code of Life.* New York: Plenum.

Mann, C. C. (1994) "War of Words Continues in Violence Research." *Science* 263: 1375.

May, R., ed. (1959) *Existence.* New York: Basic Books.

Mayr, E. (1982) *The Growth of Biological Thought.* Cambridge: Belknap Press of Harvard University.

Meir, J. P. (1991) *The Marginal Jew.* New York: Bantam Doubleday.

Mitchell, S. (1991) *The Gospel According To Jesus.* New York: Harper Perennial.

Moffat, A. S. (1994) "Microbial Mining Boosts the Environment, Bottom Line." *Science* 264: 778-79.

Nietzsche, F. W. (1880) *The AntiChrist.* Translated by W. Kaufmann (1976) in *The Portable Nietzsche.* New York: Viking Penguin. (12th Printing).

—— (1883-1885) *Thus Spoke Zarathustra.* Translated by W. Kaufmann (1976) in *The Portable Nietzsche.* New York: Viking Penguin. (12th Printing).

O'Malley, W. J. (1994) "The Moral Practice of Jesus." *America.* 170 (April 23): 8-11.

Pendick, D. (1993) "Science & Society." Reports from Mississippi Beach, Miss. International Science and Engineering Fair in *Science News,* 143: 22 (May 29) 351.

Schneider, S. H. and P. J. Boston, eds. (1992) *Scientists on Gaia*. Cambridge: Massachusetts Institute of Technology Press.

Schweitzer, A. [1906] (1968) *The Quest for the Historical Jesus*. Reprint. New York: Macmillan.

Shannon, P. (1992) "Gaia Without Mysticism." *Skeptical Inquirer*. Vol. 17, No. 1, Fall.

Shreeve, J. (1994) "Lucy, Crucial Early Human Ancester, Finally Gets A Head." *Science*. 264: 34.

Taubes, G. (1994) "Heisenberg's Heirs Exploit Loopholes in His Law." *Science*. 262: 1376-77.

"Tricks to Make DNA Beget DNA" (1994) *Science News*, 145: 22(May 28) 349.

Vignettes: Pitfalls of Evolution. (1994) Hardy quoted in "Reinventing the Future, Conversations With the World's Leading Scientists." (1994) by T. A. Bass. New York: Addison-Wesley. *Science*. 263: 1301.

Watson, J. D. (1986) *The Double Helix*. New York: Dutton.

Webster's New World Dictionary. 3rd College Edition. (1988) New York: Simon & Shuster, Inc.

Wilson, E. O. (1980) *Sociobiology*. Cambridge: Belknap Press of Harvard.

Wingerson, L. (1990) *Mapping Our Genes*. New York: Dutton.

ABOUT THE AUTHOR

Dr. Kent has an undergraduate degree from Yale University and advanced degrees in psychology, anthropology and neurology from Columbia University, the University of Southern California, and Johannes Gutenberg University.

In addition to his years in private practice, Dr. Kent served as Chief Psychologist, USAF Wiesbaden Hospital, Head of the Department of Behavioral Science, University of Southern Colorado, and Director of Mental Health, Indian Health Service at the Quechan and Cocopah Reservations. Dr. Kent is a diplomate in Clinical Psychology, a Fellow of the American Psychological Association and member of the American Association For the Advancement of Science. He is listed in *Who's Who in America* and *Who's Who in the World.*

Previous books include: *A Psychologists Answers Your Questions; Conflict Resolutions; Three Warriors Against Substance Abuse; Behind the Therapists' Notes—Fears, Feelings & Hopes.*